Dynamic Cells:
Cell Biology of the 21st Century

Dynamic Cells: Cell Biology of the 21st Century

Proceedings of the 1st Hirosaki International Forum of Medical Science, Hirosaki, Japan, 30 October 1997

Edited by:

Soroku Yagihashi
Department of Pathology
Hirosaki University School of Medicine
Hirosaki, Japan

Takashi Kachi
Department of Anatomy
Hirosaki University School of Medicine
Hirosaki, Japan

Makoto Wakui
Department of Physiology
Hirosaki University School of Medicine
Hirosaki, Japan

1998

ELSEVIER

Amsterdam – Lausanne – New York – Oxford – Shannon – Singapore – Tokyo

ELSEVIER SCIENCE B.V.
Sara Burgerhartstraat 25
P.O. Box 211, 1000 AE Amsterdam, The Netherlands

First edition 1998

Library of Congress Cataloging in Publication Data
A catalog record from the Library of Congress has been applied for.

International Congress Series No. 1166
ISBN: 0 444 50001 4

∞ The paper used in this publication meets the requirements of ANSI/NISO Z39.48-1992 (Permanence of Paper).

Printed in the Netherlands

Chairman's Address

Since being founded in 1944, the Hirosaki University School of Medicine has continually developed, establishing a university hospital, a graduate school, the Institute of Cerebrovascular Disease and the Institute for Animal Experiments. In 1994, in commemoration of our medical school's 50th aniversary, staff and alumni established "The International Exchange Fund of Hirosaki University School of Medicine" to promote academic and professional international relationships all over the world.

Using this fund, we have planned a 5-year series called "The Hirosaki International Forum". The first of that series, with the theme "Dynamic Cells: Cell Biology of the 21st Century, Cell Structure and Function, an Update", was held on October 30, 1997. We were very honored to have as guest speakers the internationally active researchers: Prof Horst Werner Korf from J.W. Goethe Institute, Germany; Prof Akira Tonosaki from Yamagata University School of Medicine, Japan; Prof Nobutaka Hirokawa from the University of Tokyo, Japan; Prof Tomas Hökfelt from the Karolinska Institute, Sweden, and Staff members of our medical school. Many persons came from all over Japan to attend. I am sure the many excellent papers presented in the forum have made a great contribtuion to the field of cell biology. The forum also fostered the development of international exchange between our medical school and others. It is our pleasure to publish this monograph to extend the contributions of the forum. We plan to continue this series of monographs during future Hirosaki International Forums.

Masahiko Endo, M.D.
Dean
Hirosaki University School of Medicine
Japan

Editors' Note

We have finally launched this memorial issue, entitled "Dynamic Cells: Cell biology of the 21st century", in connection with the first Hirosaki International Forum of Medical Science. The forum ended on October 30, 1997 in a great success which owed much to the distinguished guest speakers and our representative faculties. In this volume, contributions from world-renowned scientists were included for the expansion of our purpose to edit a useful and valuable source of modern knowledge of cell biology for the medical world. We heartily appreciate the efforts given to this issue by contributors most authoritative in the field of medical cell biology. When the Hirosaki Forum was established, the organizing committee immediately all agreed with the idea that the Forum should start with the main theme of the topics in basic medicine, in particular the progress in medical cell biology. Because recent progress in modern medicine is mostly based on the clarification of molecular mechanisms of cell dynamics and genetic alterations and their regulation, the knowledge of molecular cell biology is essential for modern physicians and even for comedical staff. We have to think about and deal with, in detail, the causes and mechanisms of diseases and directions of treatment at the cellular biological levels. To attract people to the Hirosaki Forum, we wanted the topics referred to in the Forum to be related to all fields of medicine and strongly implicated in future clinical application. Such an ambition appeared to be so demanding that the Forum would not be able to cover all the requirements. However, looking at this volume in hand, we found it needless to worry about the content. Mechanisms of cell movement and intracellular transport of materials, secretion, synthesis of proteins, reactions to the stimuli in the cell are described with completely novel thoughts. Nerve tissues, endocrine cells, intestinal secretory cells, photoreceptors, perception of hearing, bone cells, and pancreatic islets are the targets of new discoveries in modern cell science. Readers will know how the dynamic process of intracellular milieu are well-organized. We expect that this volume will not only supplement your knowledge on cell biology obtained from textbooks, but also stimulate your inspiration concerning ongoing research and medical practice. In addition, we sincerely hope that this volume stands for a milestone in the international academic activities of our medical school.

Soroku Yagihashi
Takashi Kachi
Makoto Wakui
Editors and the Organizing Committee of the 1st
Hirosaki International Forum of Medical Science

Contents

Dynamic Cells: Cell Biology of the 21st Century
S. Yagihashi, T. Kachi and M. Wakui, editors.

Plasticity in expression of neuronal messenger molecules: mechanisms and functional aspects

Tomas Hökfelt[1], Zhang Xu[1,3], Zhi-Qing David Xu[1], Tiejun Sten Shi[1], Jacquie Corness[1], Nóra Kerekes[1], Marc Landry[1,4], Kristina Holmberg[1], Malin Rydh-Rinder[2], Christian Broberger[1], Ji Ru-Rong[1] and Robert Elde[5]

[1]Department of Neuroscience, Karolinska Institutet, Stockholm; [2]Astra Pain Control Södertälje, Sweden; [3]The Fourth Military Medical University, Institute of Neurosciences, P.R. China; [4]Laboratoire de Biologie Cellulaire, Bordeaux, France; and [5]Department of Cell Biology and Neuroanatomy, University of Minnesota, Minneapolis, USA

Abstract. With immunohistochemistry and in situ hybridization methodologies we have used rat dorsal root ganglion (DRG) neurons as a model to explore the localization, expression, regulation and possible functions of peptide and peptide receptors. Many DRG neurons express peptide and peptide receptors under normal circumstances. Peripheral manipulations, in particular nerve injury, cause marked changes in expression of these molecules. Thus, the excitatory peptides substance P and calcitonin gene-related peptide (CGRP) are markedly downregulated, whereas two inhibitory peptides, galanin and neuropeptide Y (NPY), are strongly upregulated after axotomy. With regard to receptors, CCK_B receptors are hardly detectable in normal DRG neurons, but are strongly upregulated in a large proportion of DRG neurons after axotomy. The NPY Y1 receptor is present in many, mainly small DRG neurons but is not transported centrifugally and thus represents a somatic receptor. It is downregulated after axotomy. Both galanin R1 and R2 receptor mRNAs can be detected in DRG neurons normally and they are both downregulated after axotomy. The mechanisms underlying these regulatory events are only incompletely understood, but NGF (nerve growth factor) is important for maintaining normal levels of substance P and CGRP, whereas leukemia inhibitory factor is important for upregulation of galanin. Functionally, it is postulated that the regulatory events in DRG neurons aim at suppressing transmission in the dorsal horn but also to support regenerative events. Thus, galanin may act as an analgesic factor, in particular after nerve injury. The changes described above here can not be seen after peripheral inflammation, which in contrast causes a marked increase in expression of, for example, opioid peptides in local dorsal horn neurons. It is therefore our hypothesis that two pain defense systems exist, one consisting of local dorsal horn neurons releasing inhibitory peptides such as enkephalin and dynorphin, and they suppress inflammatory pain, and another one present in the DRG neurons themselves, which downregulate excitatory peptides and release inhibitory peptides to attenuate nerve injury-induced pain.

Keywords: coexistence, galanin, injury, nerve, neuropeptide Y, pain.

Introduction: basic aspects on neuropeptides

In the beginning of the 1970s some biologically active peptides were discovered

---

Address for correspondence: Tomas Hökfelt, Department of Neuroscience, Karolinska Institutet, S-171 77 Stockholm, Sweden. Tel.: +46-8-7287070. Fax: +46-8-331692.
E-mail: Tomas.Hokfelt@neuro.ki.se

and/or structurally characterized, including substance P, several releasing and inhibiting hypothalamic hormones, and the two pentapeptides methionine-and leucine-enkephalin which are endogenous ligands for opioid receptors. Subsequently further neuropeptides have been discovered (Table 1); however, it has been difficult to establish a precise physiological role for many of them. Admittedly, the roles of the hypothalamic hormones oxytocin and vasopressin as well as the hypothalamic releasing and inhibitory factors are well-established, but we still search for the significance of peptides in many other systems.

Neuropeptides can not only be grouped into families as shown in Table 1, but can also be classified according to other principles, for example, with regard to level of expression. Thus in some systems a certain peptide is expressed at high levels under normal circumstances, and in other cases its expression is low but can be increased by a certain stimulus. A third mode is the transient expression during development (see also [1]).

An important advancement in our understanding of neuropeptidergic mechanisms has been the cloning and characterization of receptors and receptor subtypes for most peptides. With a few exceptions they belong to the group of G-protein coupled receptors with seven transmembrane segments [2,3]. Also, many nonpeptide antagonists have been developed by pharmaceutical companies [3].

It should be emphasized that peptides almost always are present in neurons which already utilize another compound as its principal transmitter, for example acetylcholine, serotonin, or GABA (see [4]). Thus peptides are certainly not the sole messenger molecules in neurons, and could therefore also have roles completely different from those of the classic transmitters, for example, exerting long-term and trophic effects. In our attempts to understand peptide function we have focused on the sensory neurons in dorsal root ganglia (DRG) as a model (Fig. 1). Some of these neurons relay pain sensation and express robust levels of several peptides as well as peptide receptors. Most peptides and receptors in these neurons also show interesting changes in expression patterns in response to various types of manipulations.

Peptides and their receptors in DRG neurons

The first peptide to be described in sensory neurons was substance P which is present in around 20% of all DRG neurons. Somatostatin and calcitonin gene-related peptide (CGRP) are also found in DRGs, mainly in small neurons. CGRP is present in around 50% of all DRG neurons, whereby most substance P neurons contain CGRP (for a review see [5]). This neuron population to a large extent also expresses the neuropeptide Y (NPY) Y1 receptor [6], whereas a neurotensin receptor occupies a different subset of neurons [6].

Functionally both substance P and CGRP facilitate transmission in the dorsal horn [7–10], whereby CGRP not only acts directly on dorsal horn neurons [8,9] but also enhances substance P action [11] by inhibiting a substance P endopeptidase [12]. However, the principal transmitters in primary sensory neurons are

Table 1. Some mammalian neuropeptides and neuropeptide families.

Hypothalamic hormones
Oxytocin (9 amino acid residues, a.a.r.)
Vasopressin (9 a.a.r.)

Hypothalamic releasing and inhibiting hormones
Thyrotropin releasing hormone (TRH) (3 a.a.r.)
Luteinizing hormone releasing hormone (LHRH) (10 a.a.r.)
Somatostatin (14 a.a.r. and other forms)
Corticotropin releasing hormone (CRH) (41 a.a.r.)
Growth hormone releasing hormone (GHRH) (44 a.a.r.)

Tachykinins
Substance P (11 a.a.r.)
Neurokinin A (substance K) (10 a.a.r.)
Neurokinin B (10 a.a.r.)
Neuropeptide K (36 a.a.r.)

Opioid peptides
Met- and leu-enkephalin (5 a.a.r.)
Dynorphin (17 a.a.r. and other forms)
β-endorphin (30 a.a.r.)

NPY and related peptides
Neuropeptide tyrosine (NPY) (36 a.a.r.)
Pancreatic polypeptide (36 a.a.r)
Peptide tyrosine-tyrosine (PYY) (36 a.a.r.)

VIP-glucagon family
Vasoactive intestinal polypeptide (VIP) (28 a.a.r.)
Peptide histidine-isoleucine (PHI) (27 a.a.r.)
Pituitary adenylate cyclase activating peptide (PACAP) (27 or 38 a.a.r.)
Glucagon-like peptide-1 (GLP-1) (29 a.a.r.)

Other peptides
Cholecystokinin (CCK) (8 a.a.r. and other forms)
Neurotensin (13 a.a.r.)
Calcitonin gene-related peptide (CGRP) (α- and β-form) (37 a.a.r.)
Galanin (29 or 30 a.a.r.)
Parathyroid hormone related protein (34 or 37 a.a.r.)
Brain natriuretic peptide (32 a.a.r.)
Islet amyloid polypeptide (IAPP) or amylin (37 a.a.r)

Novel neuropeptides
Secretoneurin (33 a.a.r.)
Urocortin (40 a.a.r.; 45% sequence identity with CRH)
Nociceptin or orphanin FQ (17 a.a.r.)

not neuropeptides, but glutamate and/or ATP (see [13]), a view supported by immunohistochemical studies demonstrating glutamate-like immunoreactivity in

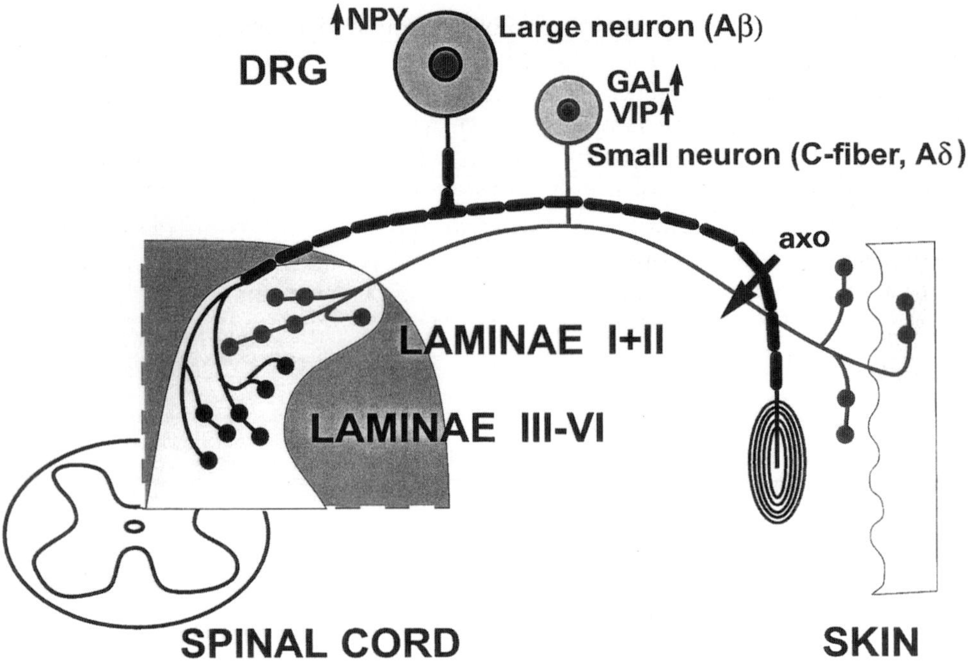

Fig. 1. Schematic illustration of small and large DRG neurons with their peripheral branches terminating in the skin and their central branches in the dorsal horn of the spinal cord. Note that axons of small neurons are thin and unmyelinated, terminating close and even within the epithelium in the skin and in the superficial laminae (I+II) of the dorsal horn. In contrast the large neurons have myelinated axons and they terminate in deeper lamina in the dorsal horn. Peptides such as galanin (GAL) and VIP are upregulated mainly in small neurons after peripheral axotomy (axo) and NPY mainly in large neurons.

many DRG neurons [14], and by the presence of several types of ATP receptors on primary sensory neurons [15,16].

Nerve injury-induced regulation of peptides in DRG neurons

After sciatic nerve transection, substance P decreases in the dorsal horn [17] which is due to a reduction in peptide synthesis in the DRG neurons [18]. Similar findings have been reported also for CGRP [19]. In contrast, vasoactive intestinal polypeptide (VIP) synthesis increases in DRG neurons after peripheral nerve injury [20], and the same is true for galanin [21,22], NPY [23,24] and PACAP (pituitary adenylate cyclase-activating peptide) [25,26]. This, however, partly occurs in different neuron populations (*cf.* Fig. 1). Thus, VIP is mainly upregulated in small DRG neurons, galanin mostly in small but also in some large neurons, NPY mostly in large and a few small neurons and PACAP in both small and large neurons. In fact, PACAP is the peptide upregulated in the highest proportion of DRG neurons, and approximately 70% of all DRG neuron profiles express this peptide a week after axotomy compared to about 50% for galanin and lower percentages for NPY and VIP. PACAP coexists with both NPY, VIP and galanin after axotomy [25], and there is a high degree of coexistence between VIP and galanin in small neurons [27,28]. Also the time course of upregulation differs between these peptides in that PACAP shows a transient expression [25], whereas galanin upregulation persists for at least several months [22]. Not only axotomy but also a number of other manipulations induce increased galanin synthesis in DRG neurons. In general there is a good correlation between increases in mRNA and peptide levels, suggesting an efficient translation.

Nerve injury-induced regulation of peptide receptors in DRG neurons

Peptide receptors are present in DRG neurons, and they are regulated by nerve injury. Under normal circumstances the CCK_B receptor is expressed at very low levels in rat DRG neurons but is strongly upregulated after peripheral nerve injury, resulting in CCK_B receptor mRNA expression in approximately 60% of all neuron profiles [29]. A neurotensin receptor has been detected in about one-third of all small DRG neuron profiles, and its synthesis is downregulated after axotomy [6]. Recently two galanin receptors have been cloned. The GAL-R1 receptor has a wide distribution in the brain [30–32], and is expressed in about 20% of all DRG neuron profiles, virtually all of which contain CGRP mRNA [33]. Axotomy induces a marked downregulation of expression of GAL-R1 mRNA, and inflammation causes a transient decrease in mRNA levels [33]. Also the GAL-R2 receptor [34,35] is expressed in a population of DRG neurons, often colocalized with CGRP [36]. Again axotomy causes a decrease in GAL-R2 mRNA levels, but there is a transient increase after inflammation [36].

With regard to sensory and, in particular, pain processing, much interest has focused on the three opioid receptors, and the μ, δ and κ receptors (MOR, DOR and KOR, respectively) have now been cloned (see [37]). Using in situ hybridization and immunohistochemistry the expression of all three receptors has been studied and found, i.a. in DRG neurons [37–43]. MOR mRNA was found in more than 50% of all DRG neurons [40], and 40% of these neurons expressed preprotachykinin (ppTK) mRNA, and almost all ppTK mRNA-positive neurons expressed MOR mRNA [40]. The levels of MOR mRNA and MOR-like immunoreactivity (LI) were downregulated after peripheral nerve injury, and this downregulation was particularly marked in monkeys [43]. This is in agreement with the downregulation of opiate receptors as revealed with ligand binding techniques [17,44,45], and may account for the well known, clinically observed insensitivity to morphine in neuropathic pain [46,47]. Also the DOR is downregulated in DRG neurons after axotomy [42], which may further contribute to opiate insensitivity.

An interesting aspect on DORs is their localization. Thus, it could be shown at the ultrastructural level that DOR-LI is present in large dense core vesicles [42,48], that is the structures storing peptides such as substance P and CGRP.

Using preembedding nanogold labeling and silver enhancement, it could be demonstrated that the receptor protein appears to be associated with the membrane of these vesicles [42,49]. We have proposed that the DOR in DRG neurons

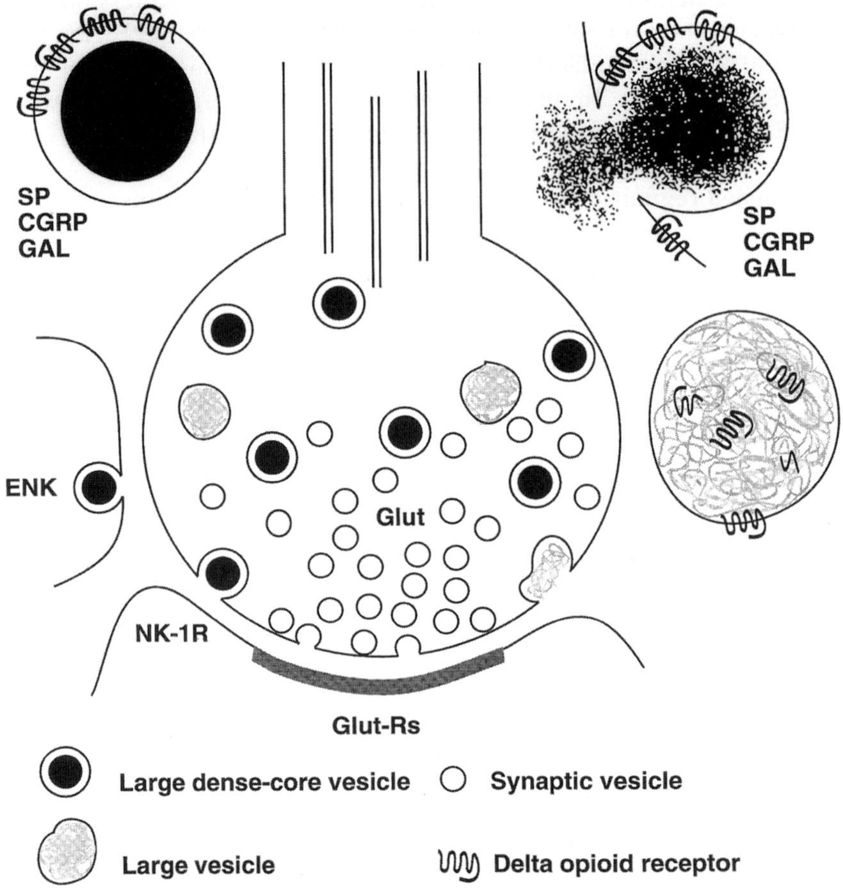

Fig. 2. Schematic illustration of our view on the localization and processing of δ-opioid receptor (R)-LI in the primary afferent terminal. Many synaptic vesicles presumably containing glutamate (Glu), and a smaller number of large dense-core vesicles (LDCVs) containing substance P (SP), CGRP and galanin (GAL) are seen. δ-Opioid-R-LI is located on the membrane of the LDCVs, and on the plasmalemma of the terminals, where LDCVs fuse with the plasmalemma during exocytotic release. Moreover, δ-opioid-R-LI is also located on the membrane and inside of large vesicles having a light, structured matrix and lacking a distinct electron-lucent halo between the membrane and the matrix. Large vesicles can be observed in the terminals and on the plasmalemma of the terminals. It is proposed that δ-opioid-Rs are sorted as a membrane protein of LDCVs at the Golgi complex and transported to the terminals. When the peptides are released from LDCVs upon nociceptive stimuli, the δ-opioid-Rs are transiently incorporated in the plasmalemma in such a way that enkephalin (ENK) released from adjacent structures can activate the δ-opioid-Rs. This process attenuates/terminates the release of SP and other peptides (see [81]). The internalized receptors are processed for recycling or degradation. (From [42]).

spans the vesicle membrane seven times with its N-terminal portion in the interior of the vesicle and its C-terminal on the outside in the cytoplasm (Fig. 2) [42,49]. We further hypothesize that upon exocytotic release, the N-terminal portion will be exposed to the external (synaptic/parasynaptic) compartment allowing a ligand, e.g., enkephalin, released from adjacent nerve endings/dendrites to activate the DOR (Fig. 2). If true, this short and transient release-associated activation of a receptor seems to represent a novel type of ligand-receptor interaction.

Species differences

The rat is the preferred species in most studies on plasticity of peptide and peptide-receptor expression in DRG neurons. However, when considering pain mechanisms and pain treatment it is important to perform comparative analyses. Focusing on the three peptides galanin, NPY and substance P, both similarities and differences have been encountered in some other species. Thus the same changes as described above in rat also occur in mice, that is, upregulation of galanin and NPY and downregulation of substance P [50]. In the guinea pig there is downregulation of substance P and a strong upregulation of NPY but only a weak upregulation of galanin [51]. The galanin upregulation in the cat is also very weak [52]. In the monkey (M. mulatta) substance P and galanin follow the patterns in the rat, whereas NPY is not upregulated [53]. Moreover, we have recently been able to analyze two human DRGs extirpated for therapeutic purposes and frozen rapidly after surgery (Landry, Åman, Dostrowski, Lozano, Zhang, Wiesenfeld-Hallin and Hökfelt, in preparation). Galanin was expressed in both ganglia showing that human DRGs can express galanin at substantial levels. Galaninergic mechanisms may, therefore, operate at the spinal level of human, and could be involved in processing of pain stimuli. This is important knowledge, since it has been reported that, whereas galanin is expressed in a large portion of the cholinergic forebrain neurons in New World monkeys, none of these neurons in the human forebrain were galanin-immunoreactive [54].

Electrophysiological studies

Recent studies in our laboratory have attempted to compare the changes in receptor expression and the sensitivity of acutely removed DRG neurons to exogenously-applied peptides using electrophysiological techniques [55]. For example, no effect of galanin could be recorded in normal DRG neurons, but after axotomy this peptide caused depolarization (Fig. 3), in agreement with an in vitro study showing that galanin has an excitatory effect on cultured DRG neurons [56]. The two studies seem to agree, since it has been proposed that DRG neurons in culture with regard to peptide phenotype represent axotomized neurons [57]. However, the results are still surprising, since, as mentioned above, both GAL-R1 and GAL-R2 mRNA are found in normal DRG neurons, but are downregulated after axotomy [33,36]. Thus, the inward current induced by galanin in axo-

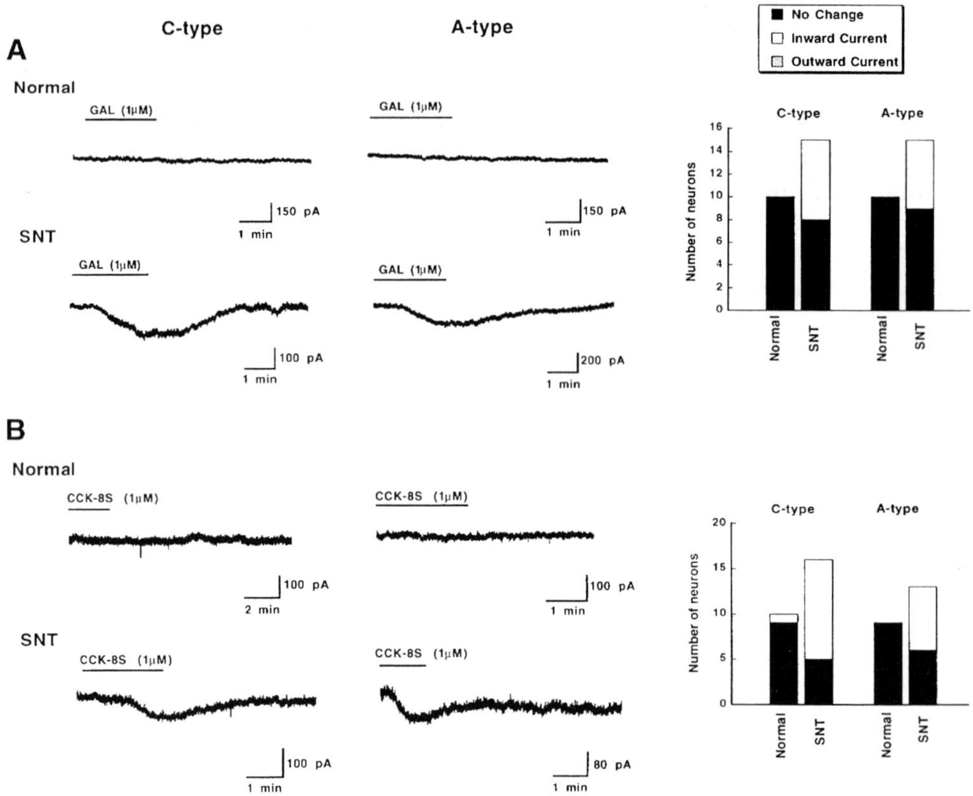

Fig. 3. **A:** Bath application of GAL does not change membrane current in a C-type (upper left, holding potential −60 mV) or A-type (upper right, 65 mV) cell from normal rats but induces inward currents in a C-type (lower left, −60 mV) and A-type (lower right, −65 mV) cell from SNT (sciatic nerve transection) rats. **B:** CCK-8S at 1 μM does not change the membrane current in either a C-type (upper left, −62 mV) or A-type (upper right, −65 mV) cell from normal rats; however, it induces an inward current in both a C-type (lower left, −62 mV) and an A-type (lower right, −65 mV) cell from SNT rats. (From [55].).

tomized DRG neurons could be mediated by a third receptor. Furthermore, we have shown that DRG neurons normally are not sensitive to sulphated CCK octapeptide, but after axotomy a large number of neurons are depolarized by CCK (Fig. 3), in good agreement with the upregulation of CCK_B receptors after axotomy [29].

Mechanisms underlying peptide regulation after nerve injury

It was previously shown that nerve growth factor (NGF) is important for the expression of both substance P and CGRP in DRG neurons [58–61]. We have analyzed to what extent NGF could be involved in the axotomy-induced upregulation of galanin in several experiments. In the first study a partial counteraction of the axotomy-induced upregulation of galanin was observed 10 days after intrathecal application of NGF via an osmotic pump [61]. In the second experiment the expression of galanin was studied in vitro in DRG cultures [57]. Approximately 30% of all DRG neurons in these cultures expressed high galanin levels, suggesting that these neurons are in an axotomized state. NGF was administered to those cultures, and a fairly high concentration (100 ng/ml) significantly reduced galanin expression [57]. In the third study, when NGF was applied locally via a capsule onto the transected sciatic nerve, no effect on galanin upregulation was recorded, although the downregulation of substance P was fully counteracted [62]. In the fourth study, there was a marked increase in galanin levels in cultured DRG neurons, when neutralization of NGF by antibodies was combined with administration of LIF (leukemia inhibitory factor) [63]. These studies suggest that NGF is involved in, but is not the only, or even the most important factor for the regulation of galanin expression after nerve injury (see below).

Growth factors seem to be important for some other peptides that are upregulated after nerve injury. For example, it has been demonstrated that both acidic fibroblast growth factor (FGF-1) and basic FGF (FGF-2) counteract the lesion-induced upregulation of NPY in DRG neurons [62]. This is in agreement with the presence of these growth factors and their receptors in DRGs. Thus, FGF-1 is expressed at high levels in a large population of DRG neurons under normal circumstances [64], and FGF-2 has a low expression under normal circumstances but is dramatically and rapidly upregulated after nerve injury, again in a large population of DRG neurons [65]. A role for FGFs in sensory neurons is strongly suggested by the fact that large numbers of DRG neurons express at least two of the FGF receptors, FGF R-1 and FGF R-2 but not FGF R-3 or FGF R-4 [66]. The FGFR-1 and -2 expressing neurons were mainly medium- or large-sized, which would be in agreement with our findings that FGF-1 and -2 mainly influence NPY expression [62], since NPY is mainly present in large neurons after axotomy [23,24].

More recently, evidence has been presented that LIF or cholinergic differentiation factor (CDF) [67,68] represents a key molecule for the expression of transmitters and peptides in sympathetic and sensory neurons, particularly after nerve injury [69—71]. Thus, the increase in galanin in the superior cervical ganglion [69] and in DRG neurons [50,72] seen after nerve injury is markedly reduced in LIF-deficient mice. Finally, as described above, LIF strongly stimulates galanin synthesis in DRG neurons in vitro, provided that NGF is neutralized [63].

Some functional aspects

The functional significance of the dramatic changes in expression of peptides and receptors, as well as of other molecules, is only incompletely understood. It has been suggested that the downregulation of substance P and CGRP may attenuate information processing in the dorsal horn, in particular related to pain, since both peptides are present in small neurons and have been shown to mainly have excitatory actions as revealed in many studies. This would fit with the idea that these plastic changes serve to induce antinociception (see [73]). The purpose of the upregulation of peptides appears more complex. Both galanin and NPY have in several instances been shown to cause inhibition, and an increased release in the dorsal horn could, therefore, further attenuate processing of nociceptive information. However, as shown in several studies, galanin and NPY can also have facilitatory effects in the dorsal horn, in addition to causing inhibition (see [73]). Furthermore, completely different types of functions should also be considered including effects on survival and regeneration. In fact, NPY has recently been shown to enhance neurite elongation [74], and in galanin knockout mice regeneration is slower than in wild-type (Wynick, personal communication). Such a role for galanin has been further strengthened by the demonstration that this peptide is highly expressed during development in several systems, including DRG [75,76], suggesting involvement in developmental processes.

We have performed studies on autotomy, a self-mutilation behavior that animals show after nerve injury and which may be related to painful sensations (see [77]), with the aim to find a relation to galaninergic mechanisms. In one study we used galanin antisense probes and in the second one a putative galanin antagonist. In the former experiment the galanin antisense probe was administered via a capsule onto the transected sciatic nerve, and galanin levels in DRG neurons were monitored by immunohistochemistry 3 days after lesion, and autotomy every day during this period [78]. There was a good correlation between the decrease in galanin-LI and autotomy, which was observed in 80% of the rats but only in 3% of the controls [78]. Also the putative galanin antagonist M35 [79] induced a marked increase in autotomy behavior, whereby all drug-treated rats showed autotomy after 10 days and only 15% in the control group [80]. These findings suggest that depletion of galanin as well as pharmacological blockade of galanin transmission in the dorsal horn increase autotomy behavior, possibly a sign of pain, in agreement with the view that the increased levels of galanin after axotomy serve to suppress pain.

Acknowledgements

This work is supported by the Swedish Medical Research Council (04X-2887), Marianne and Marcus Wallenbergs Stiftelse, Gustav Vëoch Drottning Victorias Stiftelse, Astra Pain Control AB, the Nature Science Foundation of China (Grant 39525010, 39500045), the European Commission (BMH4-CT95-0172) and by grants from the National Institute on Drug Abuse, National Institute of Health (USA).

References

1. Tohyama M. An overview of the ontogeny of neurotransmitters and neuromodulators in the central nervous system. In: Björklund A, Hökfelt T, Tohyama M (eds) Handbook of Chemical Neu-

roanatomy: Ontogeny of Transmitters and Peptides in the CNS, vol 10. Amsterdam: Elsevier Science Publishers, 1992;647–650.

2. Meyerhof W, Darlison MG, Richter D. The elucidation of neuropeptide receptors and their subtypes through the application of molecular biology. In: Hucho F (ed) New Comprehensive Biochemistry. Amsterdam: Elsevier, 1993;25:335–353.
3. Schwartz TW, Gether U, Schambye HT, Hjorth SA. Molecular mechanism of action of nonpeptide ligands for peptide receptors. Curr Pharmacol Design 1995;1:325–342.
4. Hökfelt T. Neuropeptides in perspective: the last 10 years. Neuron 1991;7:867–879.
5. Dalsgaard C-J. The sensory system. In: Björklund A, Hökfelt T, Owman C (eds) Handbook of Chemical Neuroanatomy: The Peripheral Nervous System, vol 6. Amsterdam: Elsevier Science Publishers, 1988;599–667.
6. Zhang X, Xu Z-Q, Bao L, Dagerlind Å, Hökfelt T. Complementary distribution of receptors for neurotensin and NPY in small neurons in rat lumbar DRGs and regulation of the receptors and peptides after peripheral axotomy. J Neurosci 1995;15:2733–2747.
7. Henry JL. Substance P and pain: a possible relation in afferent transmission. In: Von Euler US, Pernow B (eds) Substance P. New York: Raven Press, 1977;231–240.
8. Miletic V, Tan H. Iontophoretic application of calcitonin gene-related peptide produces a slow and prolonged excitation of neurons in cat lumbar dorsal horn. Brain Res 1988;446:169–172.
9. Ryu PD, Greber G, Murase K, Randic M. Actions of calcitonin gene-related peptide on rat spinal dorsal horn neurons. Brain Res 1988;441:357–361.
10. Woolf CJ, Wiesenfeld-Hallin Z. Substance P and calcitonin gene-related peptide synergistically modulate the gain of the nociceptive flexor withdrawal reflex in the rat. Neurosci Lett 1986;66: 226–230.
11. Wiesenfeld-Hallin Z, Hökfelt T, Lundberg JM, Forssmann WG, Reinecke M, Tschopp FA, Fischer JA. Immunoreactive calcitonin gene-related peptide and substance P coexist in sensory neurons to the spinal cord and interact in spinal behavioural responses of the rat. Neurosci Lett 1984;52:199–204.
12. Le Greves P, Nyberg F, Terenius L, Hökfelt T. Calcitonin gene-related peptide is a potent inhibitor of substance P degradation. Eur J Pharmacol 1985;115:309–311.
13. Salt TE, Hill RG. Neurotransmitter candidates of somatosensory primary afferent fibers. Neuroscience 1983;10:1083–1103.
14. De Basi S, Rustioni A. Glutamate and substance P coexist in primary afferent terminals in the superficial laminae of spinal cord. Proc Natl Acad Sci USA 1988;85:7820–7824.
15. Chen CC, Akopian AN, Sivilotti L, Colquhoun D, Burnstock G, Wood JN. A P2X purinoceptor expressed by a subset of sensory neurons. Nature 1995;377:428–431.
16. Collo G, North RA, Kawashima E, Merio-Pich E, Neidhart S, Surprenant A. Cloning of $P2X_5$ and $P2X_6$ receptors and the distribution and properties of an extended family of ATP-gated ion channels. J Neurosci 1996;16:2495–2507.
17. Jessell T, Tsunoo A, Kanazawa I, Otsuka M. Substance P: depletion in the dorsal horn of the rat spinal cord after section of the peripheral processes of primary sensory neurons. Brain Res 1979;168:247–259.
18. Nielsch V, Bisby MA, Keen P. Effect of cutting or crushing the rat sciatic nerve on synthesis of substance P by isolated L5 dorsal root ganglia. Neuropeptides 1987;10:137–145.
19. Dumoulin FL, Raivich G, Streit WJ, Kreutzberg GW. Differential regulation of calcitonin gene-related peptide (CGRP) in regenerating rat facial nucleus and dorsal root ganglion. Eur J Neurosci 1991;3:338–342.
20. Shehab SA, Atkinson ME. Vasoactive intestinal polypeptide (VIP) increases in the spinal cord after peripheral axotomy of the sciatic nerve originate from primary afferent neurons. Brain Res 1986;372:37–44.
21. Hökfelt T, Wiesenfeld-Hallin Z, Villar MJ, Melander T. Increase of galanin-like immunoreactivity in rat dorsal root ganglion cells after peripheral axotomy. Neurosci Lett 1987;83:217–220.
22. Villar MJ, Cortés R, Theodorsson E, Wiesenfeld-Hallin Z, Schalling M, Fahrenkrug J, Emson

PC, Hökfelt T. Neuropeptide expression in rat dorsal root ganglion cells and spinal cord after peripheral nerve injury with special reference to galanin. Neuroscience 1989;33:587—604.
23. Wakisaka S, Kajander KC, Bennett GJ. Increased neuropeptide Y (NPY)-like immunoreactivity in rat sensory neurons following peripheral axotomy. Neurosci Lett 1991;124:200—203.
24. Wakisaka S, Kajander KC, Bennett GJ: Effects of peripheral nerve injuries and tissue inflammation on the levels of neuropeptide Y-like immunoreactivity in rat primary afferent neurons. Brain Res 1992;598:349—352.
25. Zhang Q, Shi T-J, Ji R-R, Zhang Y-T, Sundler F, Hannibal J, Fahrenkrug J, Hökfelt T. Expression of pituitary adenylate cyclase-activating polypeptide in dorsal root ganglia following axotomy: time course and coexistence. Brain Res 1995;705:149—158.
26. Zhang Y-Z, Hannibal J, Zhao Q, Moller K, Danielsen N, Fahrenkrug J, Sundler F. Pituitary adenylate cyclase activating peptide (PACAP) expression in the rat dorsal root ganglia: up-regulation after peripheral nerve injury. Neuroscience 1996;74:1099—1110.
27. Xu X-J, Wiesenfeld-Hallin Z, Villar MJ, Fahrenkrug J, Hökfelt T. On the role of galanin, substance P and other neuropeptides in primary sensory neurons of the rat: studies on spinal reflex excitability and peripheral axotomy. Eur J Neurosci 1990;2:733—743.
28. Kashiba H, Senba E, Ueda Y, Tohyama M. Colocalized but target-unrelated expression of vasoactive intestinal polypeptide and galanin in rat dorsal root ganglion neurons after peripheral nerve crush injury. Brain Res 1992;582:47—57.
29. Zhang X, Dagerlind Å, Elde RP, Castel MN, Broberger C, Wiesenfeld-Halin Z, Hökfelt T. Marked increase in cholecystokinin B receptor messenger RNA levels in rat dorsal root ganglia after peripheral axotomy. Neuroscience 1993;57: 227—233.
30. Habert-Ortoli E, Amiranoff B, Loquet I, Laburthe M, Mayaux JF. Molecular cloning of a functonal human galanin receptor. Proc Natl Acad Sci USA 1994;91:9780—9783.
31. Burgevin MC, Loquet I, Quarteronet D, Habert-Ortoli E. Cloning, pharmacological characterization, and anatomical distribution of a rat cDNA encoding for a galanin receptor. J Mol Neurosci 1995;6:33—41.
32. Parker EM, Izzarelli DG, Nowak HP, Mahle CD, Iben LG, Wang J, Goldstein ME. Cloning and characterization of rat GALR1 galanin receptor from RIN14B insulinoma cells. Molec Brain Res 1995;34:178—189.
33. Xu Z-Q, Shi T-J, Landry M, Hökfelt T. Evidence for galanin receptors in primary sensory neurones and effect of axotomy and inflammation. Neuro Rep 1996;8:237—242.
34. Howard AD, Tan C, Shiao L-L, Palyha OC, McKee KK, Weinberg DH, Geighner SC, Cascieri MA, Smith RG, Van der Ploeg LHT, Sullivan KA. Molecular cloning and characterization of a new receptor for galanin. FEBS Lett 1997;405:285—290.
35. Fathi Z, Cunningham AM, Iben LG, Battaglino PB, Ward SA, Nichol KA, Pine KA, Wang J, Goldstein ME, Iismaa TP, Antal Zimanyi I. Cloning, pharmacological characterization and distribution of a novel galanin receptor. Molec Brain Res 1997;51:49—59.
36. Shi T-JS, Zhang X, Holmberg K, Xu Z-QD, Hökfelt T. Expression and regulation of galanin-R2 receptors in rat primary sensory neurons: effect of axotomy and inflammation. Neurosci Lett 1997;237:1—4.
37. Minami M, Satoh M. Molecular biology of the opioid receptors: structures, functions and distribution. Neurosci Res 1995;23:121—145.
38. Dado RJ, Law PY, Loh HH, Elde R. Immunofluorescent identification of a delta (δ)-opioid receptor on primary afferent nerve terminals. Neuro Rep 1993;5:341—344.
39. Maekawa K, Minami M, Yabuuchi K, Toya T, Katao Y, Hosoi Y, Onogi T, Satoh M. In situ hybridization of μ- and κ-opioid receptor mRNAs in the rat spinal cord and dorsal root ganglia. Neurosci Lett 1994;168:97—100.
40. Minami M, Maekawa K, Yabuuchi K, Satoh M. Double in situ hybridization study on coexistence of μ- δ- and κ-opioid receptor mRNAs with preprotachykinin A mRNA in the rat dorsal root ganglia. Molec Brain Res 1995;30:203—210.
41. Ji R-R, Zhang Q, Law P-Y, Low HH, Elde R, Hökfelt T. Expression of μ- δ- and κ-opioid recep-

tor-like immunoreactivities in rat dorsal root ganglia after carrageenan-induced inflammation. J Neurosci 1995;15:8156—8166.

42. Zhang X, Bao L, Arvidsson U, Elde R, Hökfelt T. Localization and regulation of the delta-opioid receptor in dorsal root ganglia and spinal cord of the rat and monkey: evidence for association with the membrane of large dense-core vesicles. Neuroscience 1998;82:1225—1242.
43. Zhang X, Bao L, Shi T-J, Ju G, Elde R, Hökfelt T. Downregulation of μ opiate receptors in rat and monkey dorsal root ganglion neurons and spinal cord after peripheral axotomy. Neuroscience 1998;82:1225—1242.
44. Fields HL, Emson PC, Leigh BK, Gilbert RF, Iversen LL. Multiple opiate receptors on primary afferent fibers. Nature 1980;284:351—353.
45. Besse D, Lombard MC, Perrot S, Besson JM. Regulation of opioid binding sites in the superficial dorsal horn of the rat spinal cord following loose ligation of the sciatic nerve: comparison with sciatic nerve section and lumbar dorsal rhizotomy. Neuroscience 1992;50:921—933.
46. Arnér LD, Meyerson BA. Lack of analgesic effect of opioids on neuropathic and idiopathic forms of pain. Pain 1988;33:11—23.
47. Kupers RC, Konings H, Adriensen H, Gybels JM. Morphine differentially affects the sensory and affective pain ratings in neurogenic and idiopathic forms of pain. Pain 1991;47:5—12.
48. Cheng PY, Svingos AL, Wang H, Clarke CL, Jebab S, Beczkowska IW, Inturrisi CE, Pickel VM. Ultrastructural immunolabeling shows prominent presynaptic vesicular localization of δ-opioid receptor within both enkephalin- and nonenkephalin-containing axon terminals in the superficial layers of the rat cervical spinal cord. J Neurosci 1995;15:5976—5988.
49. Elde R, Arvidsson U, Riedl M, Vulchanova L, Lee J-H, Dado R, Nakano A, Chakrabarti S, Zhang X, Loh HH, Law P, Hökfelt T, Wessendorf M. Distribution of neuropeptide receptors: new views of peptidergic neurotransmission made possible by antibodies to opioid receptors. Ann New York Acad Sci 1995;757:390—404.
50. Corness J, Shi T-J, Xu Z-Q, Brulet P, Hökfelt T. Influence of leukemia inhibitory factor on galanin/GMAP and neuropeptide Y expression in mouse primary sensory neurons after axotomy. Exp Brain Res 1996;112:79—88.
51. Rydh-Rinder M, Holmberg K, Elfvin L-G, Wiesenfeld-Hallin Z, Hökfelt T. Effects of peripheral axotomy on neuropeptides and nitric oxide synthase in dorsal root ganglia and spinal cord of the guinea pig: an immunohistochemical study. Brain Res 1996;707:180—188.
52. Arvidsson U, Ulfhake B, Cullheim S, Bergstrand A, Theodorsson E, Hökfelt T. Distribution of ^{125}I-galanin and its coexistence with 5-hydroxytryptamine in the cat spinal cord: biochemical, histochemical and experimental studies at the light and electron microscopic level. J Comp Neurol 1991;308:115—138.
53. Zhang X, Ju G, Elde R, Hökfelt T. Effect of peripheral nerve cut on neuropeptides in dorsal root ganglia and the spinal cord of monkey with special reference to galanin. J Neurocytol 1993;22: 342—381.
54. Kordower JH, Mufson EJ. Galanin-like immunoreactivity within the primate basal forebrain: differential staining patterns between humans and monkeys. J Comput Neurol 1990;294: 281—292.
55. Xu Z-QD, Zhang X, Grillner S, Hökfelt T. Electrophysiological studies on rat dorsal root ganglion neurons after peripheral axotomy: changes in responses to neuropeptides. Proc Natl Acad Sci USA 1997;94:13262—13266.
56. Puttick RM, Pinnock RD, Woodruff GN. Galanin-induced membrane depolarization of neonatal rat cultured dorsal root ganglion cells. Eur J Pharmacol 1994;254:303—306.
57. Kerekes N, Landry M, Rydh-Rinder M, Hökfelt T. The effect of NGF, BDNF and bFGF on expression of galanin in cultured rat dorsal root ganglia. Brain Res 1997;754:131—141.
58. Fitzgerald M, Wall PD, Goedert M, Emson C. Nerve growth factor counteracts the neurophysiological and neurochemical effects of chronic sciatic nerve section. Brain Res 1985;332:131—141.
59. Lindsay RM, Harmar AJ. Nerve growth factor regulates expression of neuropeptide genes in adult sensory neurons. Nature 1989;337:362—364.

60. Mulderry PK. Neuropeptide expression by newborn and adult rat sensory neurons in culture: effects of nerve growth factor and other neurotrophic factors. Neuroscience 1994;59:673–688.
61. Verge VMK, Richardson PM, Wiesenfeld-Hallin Z, Hökfelt T. Differential influence of nerve growth factor on neuropeptide expression in vivo: a novel role in peptide suppression in adult sensory neurons. J Neurosci 1995;15:2081–2096.
62. Ji R-R, Zhang Q, Pettersson RF, Hökfelt T. aFGF, bFGF and NGF differentially regulated neuropeptide expression in dorsal root ganglia after axotomy and induce autotomy. Reg Pept 1996; 66:179–189.
63. Corness J, Stevens B, Fields RD, Hökfelt T. NGF and LIF both regulate galanin gene expression in primary DRG cultures. Neuro Rep 1998;9:1533–1536.
64. Elde R, Cao YH, Cintra A, Brelje TC, Pelto-Huikko M, Junttila T, Fuxe K, Pettersson RF, Hökfelt T. Prominent expression of acidic fibroblast growth factor in motor and sensory neurons. Neuron 1991;7:1–20.
65. Ji R-R, Zhang Q, Zhang X, Piehl F, Reilly T, Pettersson RF, Hökfelt T. Prominent expression of bGF in dorsal root ganglia after axotomy. Eur J Neurosci 1995;7:2458–2468.
66. Oellig C, Pirvola U, Taylor L, Elde R, Hökfelt T, Pettersson RF. Acidic FGF and FGF receptors are specially expressed in neurons of developing and adult rat dorsal root ganglia. Eur J Neurosci 1995;7:863–874.
67. Fukada K. Purification and partial characterization of a cholinergic neuronal differentiation factor. Proc Natl Acad Sci USA 1985;82:8795–8799.
68. Yamamori T, Fukada K, Aebersold R, Korsching S, Fann MJ, Patterson PH. The cholinergic neuronal differentiation factor from heart cells is identical to leukemia inhibitory factor. Science 1989;246:1412–1416.
69. Rao MS, Sun Y, Escary JL, Perreau J, Tresser S, Patterson PH, Zigmond RE, Brulet P, Landis SC. Leukemia inhibitory factor mediates an injury response but not a target-directed developmental transmitter switch in sympathetic neurons. Neuron 1993;11:1175–1185.
70. Banner LR, Patterson PH. Major changes in the expression of the mRNAs for cholinergic differentiation factor/leukemia inhibitory factor and its receptor after injury to adult peripheral nerves and ganglia. Proc Natl Acad Sci USA 1994;91:7109–7113.
71. Patterson PH. Leukemia inhibitory factor, a cytokine at the interface between neurobiology and immunology. Proc Natl Acad Sci USA 1994;91:7833–7835.
72. Sun Y, Zigmond RE. Leukaemia inhibitory factor induced in the sciatic nerve after axotomy is involved in the induction of galanin in sensory neurons. Eur J Neurosci 1996;8:2213–2220.
73. Wiesenfeld-Hallin Z, Bartfai T, Hökfelt T. Galanin in sensory neurons in the spinal cord. Front Neuro Endocrinol 1992;13:319–343.
74. White DM, Mansfield K. Vasoactive intestinal polypeptide and neuropeptide Y act indirectly to increase neurite outgrowth of dissociated dorsal root ganglion cells. Neuroscience 1996;73: 881–887.
75. Marti E, Gibson SJ, Polak JM, Facer P, Springball DR, Aswegen GV, Aitchison M, Koltzenburg M. Ontogeny of peptide- and amine-containing neurons in motor sensory, and autonomic regions of rat and human spinal cord, dorsal root ganglia, and rat skin. J Comp Neurol 1987; 266:332–359.
76. Xu Z-Q, Shi T-J, Hökfelt T. Expression of galanin and a galanin receptor in several sensory systems and bone anlage of rat embryos. Proc Natl Acad Sci USA 1996;93:14901–14905.
77. Devor M. Sensory basis of autotomy in rats. Pain 1991;45:109–110.
78. Ji R-R, Zhang Q, Bedecs K, Arvidsson J, Zhang X, Xu XJ, Wiesenfeld-Hallin Z, Bartfai T, Hökfelt T. Galanin antisense oligonucleotides reduce galanin levels in dorsal root ganglia and induce autotomy in rats after axotomy. Proc Natl Acad Sci USA 1994;91:12540–12543.
79. Langel Ü, Land T, Bartfai T. Design of chimeric peptide ligands to galanin receptors and substance P receptors. Int J Pept Protein Res 1992;39:516–522.
80. Verge VMK, Xu X-J, Langel Ü, Hökfelt T, Wiesenfeld-Hallin Z, Bartfai T. Evidence for an endogenous inhibitory control on autotomy, a behavioral sign of neuropathic pain, by galanin in the

rat after sciatic nerve section: demonstrated by chronic intrathecal infusion of M-35, a newly developed high affinity galanin receptor antagonist. Neurosci Lett 1993;149:193–197.
81. Jessell TM, Iversen LL. Opiate analgesics inhibit substance P release from rat trigeminal nucleus. Nature 1977;268:549–551.

Dynamic Cells: Cell Biology of the 21st Century
S. Yagihashi, T. Kachi and M. Wakui, editors.

Kinesin superfamily proteins (KIFs) and the mechanism of organelle transport

Nobutaka Hirokawa
Department of Cell Biology and Anatomy, Graduate School of Medicine, University of Tokyo, Tokyo, Japan

Abstract. Cells transport and sort various proteins and lipids following synthesis to the correct destinations at appropriate velocities. These proteins and lipids are transported mostly in membranous organelles and protein complexes. This intracellular transport mechanism is thus fundamental to cellular morphogenesis and functioning. Microtubules serve as a rail on which microtubule-associated motor proteins convey their cargoes.

Recently, molecular cell biological and molecular genetics approaches have uncovered more than 20 kinesin superfamily proteins playing important roles in the transport process. This review focuses on the molecular mechanism of organelle transport in cells and describes these kinesin superfamily proteins mainly based on studies performed in our laboratory.

Keywords: microtubules, molecular motor, molecular cell biology, neuron.

Introduction

Many types of cells develop polarized structures which are very important to the functioning of the cells. Among these cells, neurons and epithelial cells are typical. The neuron is composed of dendrites, a cell body and a long axon along the direction of impulse propagation. Interestingly, because of the lack of protein synthesis machinery in the axon, all the proteins required in the axon and synaptic terminal have to be transported down the axon following synthesis in the cell body. Most proteins are conveyed as membranous organelles or protein complexes. In this sense, organelle transport in the axon is fundamentally important for neuronal morphogenesis and functioning [1,2]. Because similar mechanisms are observed in other cells, the neuron serves as a good model system to study the general mechanisms of organelle transport [1,2]. Epithelial cells also develop polarized structures that are apical and basolateral regions. Certain proteins are transported and sorted specifically to the apical vs. the basolateral membrane.

Microtubules serve as rails for the transport of organelles and are organized in a regular manner in these cells. In nerve axons the microtubules are arranged longitudinally with the plus end (fast growing end) pointing distally, while in epithelial cells microtubules are organized longitudinally with the plus end point-

Address for correspondence: Nobutaka Hirokawa, Department of Cell Biology and Anatomy, Graduate School of Medicine, University of Tokyo, Hongo 7-3-1, Tokyo, Japan. Tel.: +81-3-3812-2111 ext. 3326. Fax: +81-3-5802-8646. E-mail: hirokawa@m.u-tokyo.ac.jp ing basally. In most other cells such as fibroblasts and macrophages, microtubules radiate from the cell center with the plus ends pointing toward the periphery. In these cells, various organelles are transported via the microtubules using microtubule-associated motor proteins.

Early light microscopical studies on living nerve axons and biochemical studies on axonal transport revealed that certain groups of proteins move as membranous organelles by fast flow [3]. Electron microscopy (EM) studies have clearly demonstrated the presence of short cross bridge structures between membranous organelles and microtubules and suggested that these cross bridges are candidates for microtubule-associated motor proteins conveying the membranous organelles along microtubules [4] (Fig. 1). Video-enhanced contrast-differential interference microscopy combined with biochemical analyses revealed the presence of a microtubule plus end directed motor protein, kinesin, a microtubule-activated ATPase of 380 kDa composed of two heavy chains, KHC (120 kDa) and two light chains, KLC (64 kDa) [5—7]. Electron microscopy combined with the decoration with monoclonal antibodies against KHC and KLC clearly demonstrated that kinesin is an 80-nm rod-like molecule composed of two globular heads (10 nm in diameter), a stalk and fan-like end and that the globular heads which bind to microtubules are composed of heavy chains (Fig. 4) (kinesin) [8,9]. The KLCs were shown to comprise the fan-like end [8]. *Drosophila* KHC cDNA was reported to encode a protein of 975 aa which consists of the N terminal $\sim$350 aa forming the motor domain binding to microtubules, an α-helical coiled-coil-rich stalk domain where two heavy chains dimerize, and a tail domain [10].

In terms of the functions of kinesin, light and electron microscopic immunocytochemistry of fibroblasts and ligated nerve axons, in which anterogradely transported membranous organelles accumulate at the proximal region while retrogradely transported organelles accumulate at the distal region, clearly demonstrated that kinesin was predominantly associated with the membranous organelles transported anterogradely, thus proving kinesin is an anterograde motor for membranous organelle transport in nerve axons and other cells [11,12].

Two years after the identification of kinesin, another microtubule-activated ATPase, brain dynein or cytoplasmic dynein, was discovered. Cytoplasmic dynein is composed of two heavy chains of 530 kDa each, three intermediate chains of 74 kDa each, and four light intermediate chains of 55—60 kDa each, and moves along a microtubule from the plus end to the minus end, making it a good candidate for a motor for retrograde axonal transport [13—16]. Immunocytochemical analyses of ligated nerve axons suggested that cytoplasmic dyneins are transported anterogradely by membranous organelles and reassociate with retrogradely transported membranous organelles near the nerve terminals and then convey the organelles to the cell body [17]. Thus, two microtubule-associated motor proteins with distinct directionality have been identified.

Various types of membranous organelles, such as precursors of synaptic vesicles, precursors of synaptic plasma membranes containing proteins important

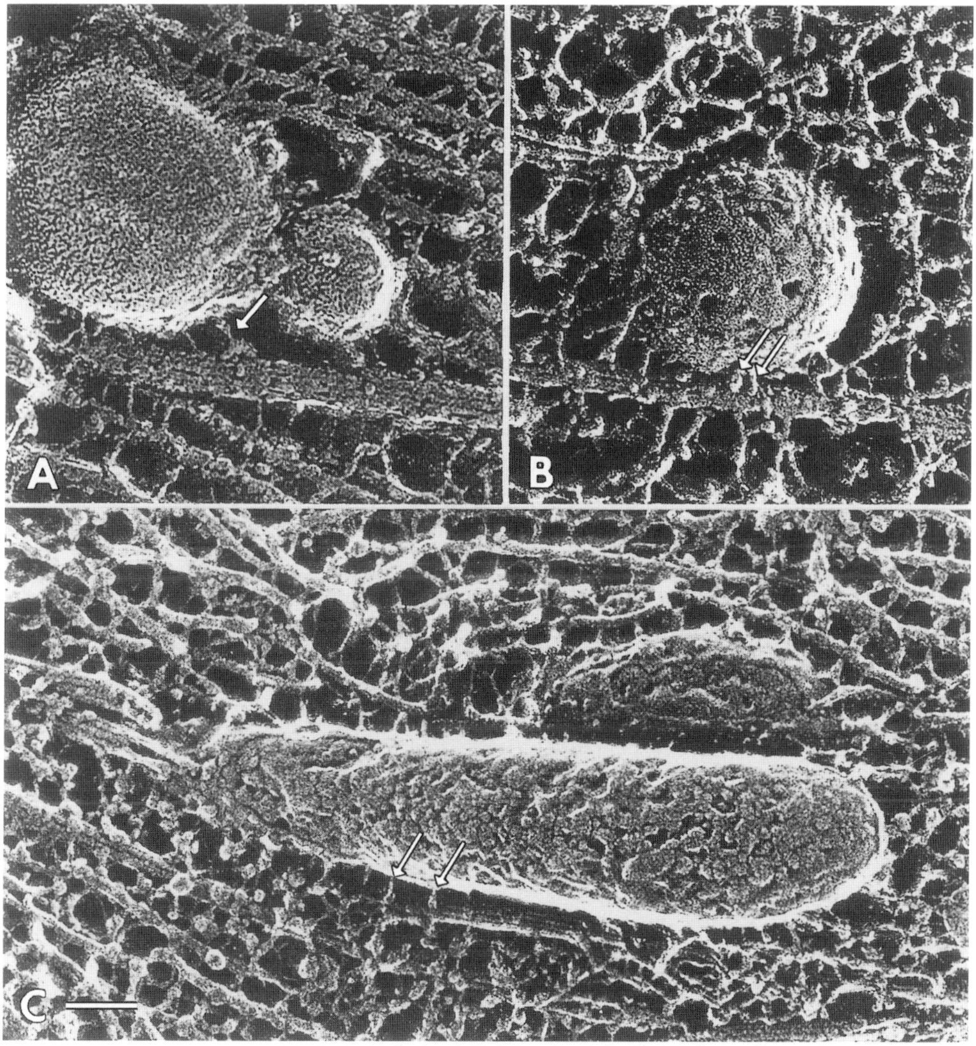

Fig. 1. Quick frozen, deep-etched axons in which various kinds of crossbridges (arrows) are identified between membranous organelles and microtubules. **A:** A crossbridge (arrow) composed of two heads and a longer stalk (~ 25 nm). **B:** Only globular heads (arrows) are recognizable between a membrous organelle and a microtubule. **C:** Crossbridges (arrows) between mitochondria and a microtubule are short and have small heads. Bar, 50 nm.

for vesicle docking such as syntaxin and SNAP 25, precursors of axonal membranes carrying Na^+ and K^+ channels, and mitochondria, are transported anterogradely, whereas multivesicular bodies, endosomes conveying neurotrophic factor receptors, and mitochondria are transported retrogradely, with both types of transport occurring at distinct velocities. Furthermore, several structurally distinct crossbridges have been observed between membranous organelles and microtubules [1]. These data strongly suggest the existence of as yet unidentified motor proteins associated with microtubules [1,2]. At that time studies of mitosis or meiosis in various metazoa such as fungus, yeast, and *Drosophila* identified putative motor proteins that have a kinesin-like motor domain containing ATP-binding and MT-binding consensus sequences [18—21]. From this background we predicted the existence of as yet unidentified microtubule-associated molecular motors functioning in organelle transport and carried out searches for these proteins.

We systematically searched the coding of kinesin superfamily genes for proteins containing ATP-binding and microtubule-binding consensus sequences from a mouse brain cDNA library enriched with cDNAs encoding proteins related to organelle transport. The library was considered to contain a small population of cDNAs encoding proteins relating to cell division because neurons are predominant in the brain and are postmitotic cells [22]. This approach led to the discovery of an initial eleven new kinesin superfamily proteins (KIFs) (KIF1A, KIF1B, KIF2, KIF3A, KIF3B, KIF4, KIF5A, KIF5C, KIFC1, KIFC2, and KIFC3) (Figs. 2 and 3) [1,2].

In this review I will describe well-characterized members of the kinesin superfamily proteins functioning in organelle transport and will focus on how they contribute to organelle transport in cells.

The kinesin superfamily proteins

According to the position of the motor domain three types of kinesin superfamily proteins have been identified: NH_2-terminal motor domain type, middle motor domain type, and COOH-terminal motor domain type.

KIF1A

KIF1A is an NH_2-terminal motor domain type monomeric anterograde motor for the transport of a subtype of synaptic vesicle precursor.

KIF1A consists of 1,695 amino acids (Mr. 191,710) [22,23], is an N-terminal motor domain type KIF and a mouse homologue of Unc-104 [24,25]. In contrast with KHC, which has a long coiled-coil stalk domain, neither a long coiled-coil stretch nor a long α-helical stretch was predicted to be present in KIF1A (Figs. 2 and 3). Low angle rotary shadowing EM revealed KIF1A to be a globular molecule with a diameter of 14 nm (Fig. 4). Differential light scattering analysis and sucrose velocity gradient centrifugation of purified KIF1A and native PAGE of native and recombinant KIF1A demonstrated that KIF1A exists as a monomer [23]. KIF1A is expressed specifically in neurons and is enriched in axons. KIF1A is the fastest moving microtubule plus end directed motor protein in mammals (1.2—1.5 μm/s in vitro) and was shown to be a motor for anterograde transport in axons. The cargo of KIF1A was identified by the double immunofluorescence microscopy analysis of ligated peripheral nerves using anti-KIF1A

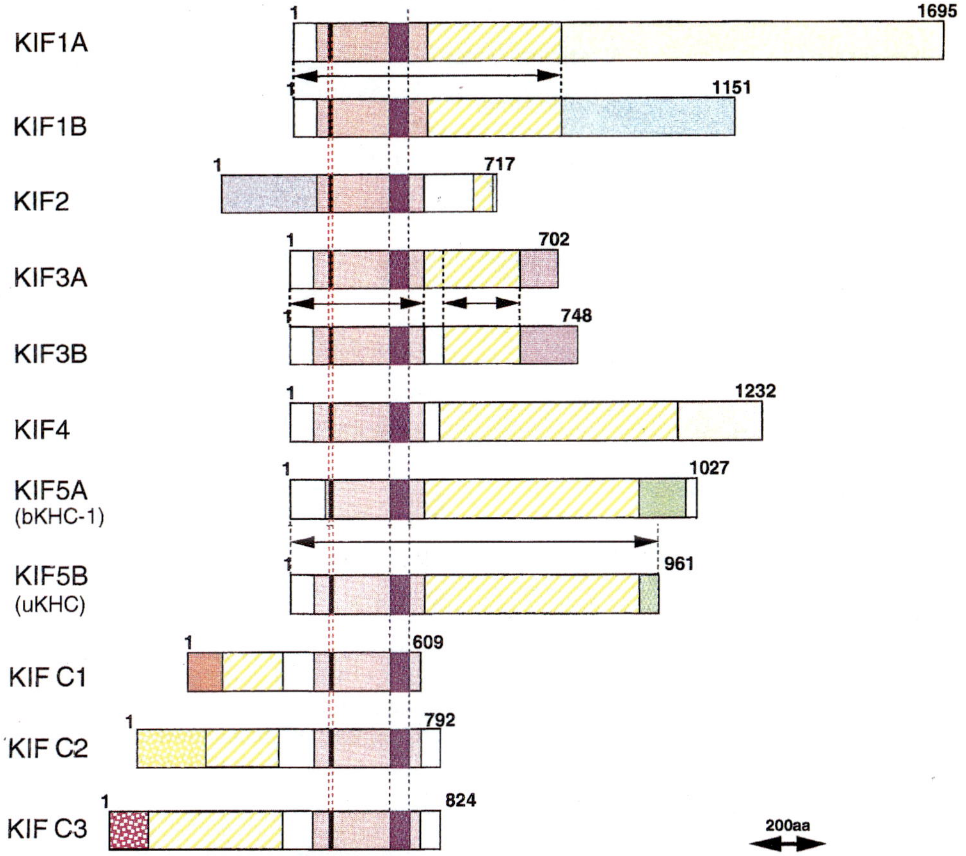

Fig. 2. Structure of cDNAs for murine kinesin superfamily proteins (KIFs). Pink: motor domain. Red: ATP-binding consensus sequence. Purple: microtubule-binding consensus sequence. KIF1A, KIF1B, KIF3A, KIF3B, KIF4, KIF5 (brain kinesin (bKHC)) and uKHC (ubiquitous kinesin) are N-terminal motors, whereas KIF2 is a middle-motor domain type. KIFC1, KIFC2 and KIFC3 are C-terminal-motor-domain types. Arrows between pairs of motor proteins indicate regions sharing strong homology.

antibody and antibodies against various kinds of proteins conveyed in the axon, and by immunoprecipitation of organelles associated with KIF1A in the axoplasm. The cargo was shown to be a subset of precursors for synaptic vesicles containing synaptophysin, synaptotagmin and Rab3A. Plasma membrane proteins important for the docking of synaptic vesicles, such as SNAP25 and syntaxin, were not contained in, and other motors such as kinesin and KIF3 were not associated with, the cargo of KIF1A [23]. Interestingly, the cargo of KIF1A did not contain SV2, a synaptic vesicle protein and transmitter transporter [23].

Analysis of KIF1A knockout mice showing sensory and motor disturbances revealed a marked reduction in the density of synaptic terminals and the density

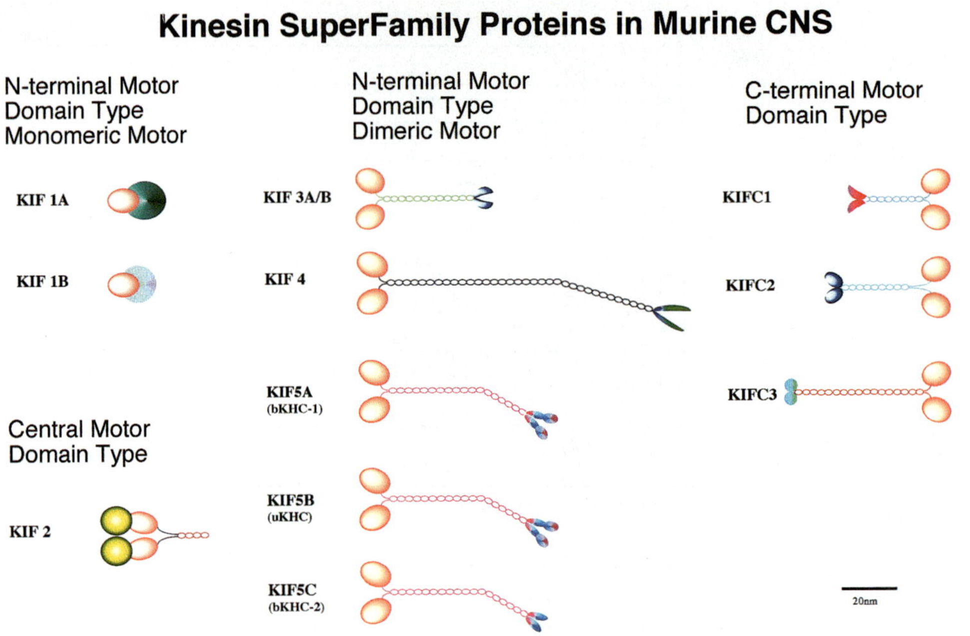

Fig. 3. Schematic drawing of KIF members in the mouse nervous system. There are various distinct types of KIFs: amino-terminal motor domain types (KIF1A, KIH1B, KIF3A/B, KIF4, KIF5A, KIF5B and KIF5C), middle-motor domain types (KIF2), and carboxy-terminal motor domain types (KIFC1, KIF2 and KIFC3), monomeric types (KIF1A and KIF1B), homodimeric types (KIF2, KIF4, KIF5A, KIF5B, KIF5C, KIFC1, KIFC2 and KIFC3) and heterodimeric types (KIF3A/B).

of synaptic vesicles in nerve terminals and ectopic accumulation of clear vesicles in nerve cell bodies. In addition, focal neuronal cell death and secondary degeneration of axons were observed in the central nervous system. The primary culture of neurons of mutants revealed prominent neuronal cell death in vitro [26]. Thus, KIF1A is a unique monomeric anterograde motor for the transport of a subset of synaptic vesicle precursors, and plays a very important role in neuronal function and neuronal survival (Fig. 5).

KIF1B

KIF1B is an $NH2_2$-terminal motor domain type monomeric anterograde motor for the transport of mitochondria.

KIF1B consists of 1150 amino acid residues (Mr 130,268) [27]. It contains neither a long coiled-coil stretch nor a long α-helical stretch and was suggested to be a monomeric motor by physicochemical analysis (Figs. 2 and 3). EM revealed that KIF1B is composed of a single globule (13–16 nm in diameter) sometimes with a short tail (Fig. 4). KIF1B exhibits a microtubule plus end directed motility ($\sim$0.66 μm/s) and is expressed in differentiated neurons, but

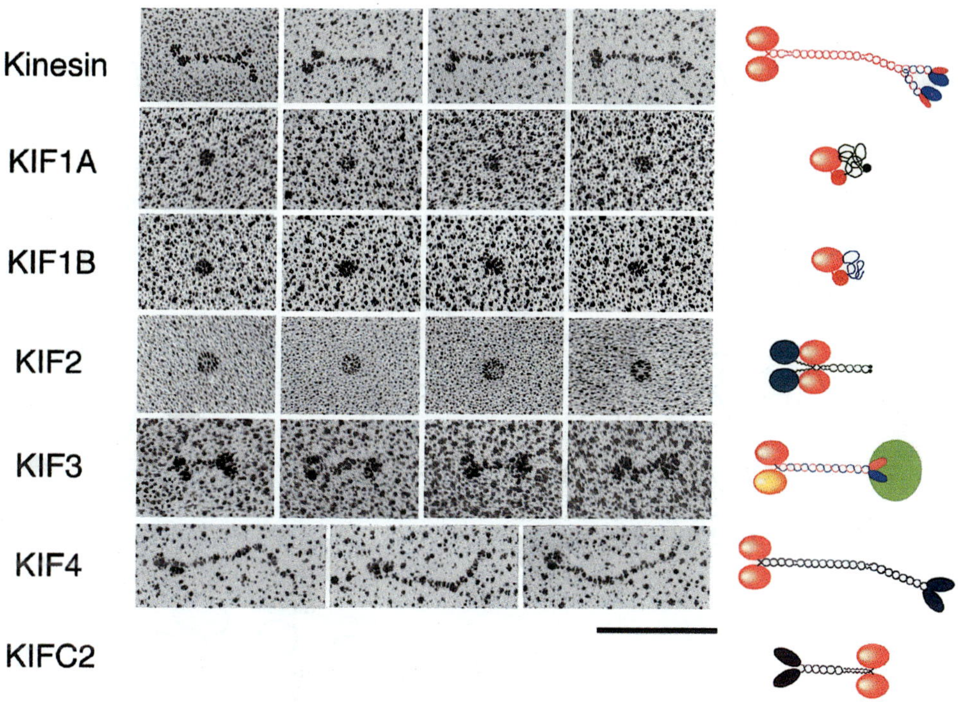

Fig. 4. Left: Panels of main members of KIFs functioning in organelle transport, as observed by low-angle rotary shadowing EM. Scale bar, 100 nm. Right: Schematic illustrations of the same KIFs, based on EM studies or predicted from analysis of primary structures.

ubiquitously in various tissues. Immunocytochemically, KIF1B was colocalized with mitochondria in vivo. Furthermore, a subcellular fractionation study showed that KIF1B was concentrated in the mitochondrial fraction and that purified KIF1B could transport mitochondria along microtubules in vitro. Thus, it was strongly suggested that KIF1B functions as a monomeric motor for the anterograde transport of mitochondria (Fig. 5) [27].

KIF3

KIF3 is an NH_2-terminal motor domain type heterodimeric motor for anterograde transport of vesicles distinct from synaptic vesicle precursor.

KIF3 is composed of a KIF3A·KIF3B heterodimer and KAP3 (kinesin superfamily associated protein 3), thus forming a heterotrimer complex (stoichiometry is KIF3A: KIF3B: KAP3, 1:1:1) (Figs. 2—4) [22,28—30]. KIF3A (701 aa, Mr 80,167) and KIF3B (747 aa, Mr 85,281) are both NH_2-terminal motor domain type KIFs, the motor domains of which are homologous while the other regions are different. Secondary structure prediction indicates that both KIF3A and KIF3B have an NH_2-terminal motor domain (globular head), a central domain

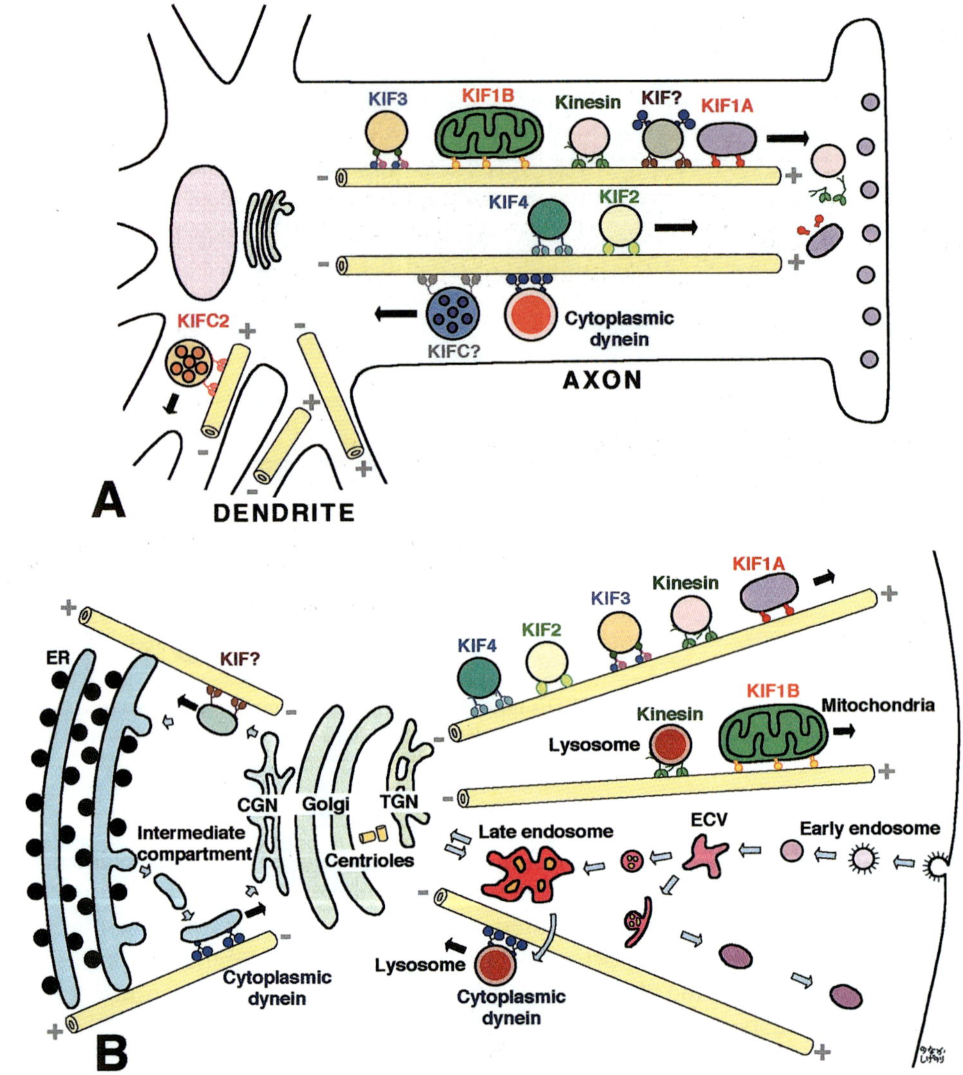

Fig. 5. Scheme of KIFs and cytoplasmic dynein and their cargo organelles in nerve axons (**A**) and in cells in general (**B**). In neurons, KIF2 and KIF4 work mostly in juvenile stages. In **B**, neuron-specific KIFs and ubiquitous KIFs are drawn in the same cell. CGN: *cis*-Golgi network. ECV: endosomal carrier vesicle. Black arrows indicate the direction of transport.

(rod α-helix domain) and a COOH-terminal domain (small tail globular domain) (Figs. 2 and 3) [22,28,29]. Immunoprecipitation of native KIF3A and KIF3B and in vitro reconstruction experiments in which KIF3A or KIF3B alone and KIF3A and KIF3B together were expressed, confirmed that KIF3A and KIF3B form a heterodimer 50 nm long consisting of two globular heads (10 nm in diameter), a stalk and tail domain (Fig. 4). The KIF3A/3B heterodimer exhibited plus end-directed microtubule sliding activity at a velocity of ~0.3 μm/s [29]. KIF3A/3B is abundantly expressed in nerve tissue, while it is also expressed ubiquitously in other tissues in lower amounts. In vivo immunocytochemical analysis of ligated nerve axons confirmed that KIF3 functions as a motor for the anterograde transport of membranous organelles. Subcellular fractionation and immunoprecipitation by anti-KIF3 antibody-conjugated beads combined with EM revealed that the cargo of KIF3 is membranous organelles, 90–160 nm in diameter, distinct from synaptic vesicle precursors and from vesicles carried by other motors such as kinesin and KIF2 [29]. The KIF3A/3B heterodimer is associated with the protein KAP3 [29]. Two isoforms of KAP3 exist (KAP3A (793 aa) and KAP3B (772 aa)), generated by alternative splicing in a single gene in the region coding for the carboxyl terminal region [30]. KAP3 is a novel protein and secondary structure prediction indicated that almost the entire KAP3 molecule is α-helical and contains armadillo repeats [30].

A reconstruction study in Sf9 cells revealed that KAP3 is a globular protein (~11 nm in diameter) that binds to the tail domain of KIF3A/3B (Fig. 4). Because KAP3 does not affect the ATPase activity of KIF3A/3B and because KAP3 is associated with the membrane-bound form of KIF3A/3B and with the tail of KIF3A/3B, KAP3 is thought to regulate the membrane binding of the KIF3 heterodimer [30]. KIF3 is a heterotrimeric motor composed of KIF3A, 3B and KAP3 and serves as an anterograde motor for transport of vesicles distinct from synaptic vesicle precursors (Fig. 5).

KIF4

KIF4 is an NH_2-terminal motor domain type dimeric motor for anterograde transport of vesicles in juvenile neurons.

KIF4 (1,231 aa, Mr 139,550) is an NH_2-terminal motor domain type KIF [31]. KIF4 (a 116-nm rod with two globular heads and a tail as determined by EM) consists of three domains: an NH_2-terminal globular motor domain, a central α-helical stalk domain to which KIF4 dimerizes, and a C-terminal tail domain (Figs. 2–4). KIF4 forms a homodimer and moves along microtubules towards the plus end at a velocity of 0.2 μm/s [31].

KIF4 mRNA is expressed abundantly in juvenile tissues including differentiated young neurons, while its expression is considerably decreased in adult mice, except in the spleen. Immunocytochemical studies revealed that KIF4 is colocalized with membranous organelles both in the growth cones of differentiated neurons and in the cytoplasm of cultured fibroblasts. During the mitotic phase of the cell cycle, KIF4 appears to colocalize with membranous organelles in the mitotic spindle. Hence, KIF4 is a microtubule plus end directed anterograde motor for the transport of a certain group of membranous organelles in juvenile neurons and other cells [31] (Fig. 5). Because KIF4 contains a DNA-binding domain similar to that of chromokinesin KIF4 may have an additional function such as the transport of mRNA in certain kinds and developmental stages of cells.

KIF2

KIF2 is a middle motor domain type anterograde motor for transport of vesicle important for juvenile neuron.

KIF2 (716 aa, Mr 80,945) is a middle motor domain type KIF [22,32]. The N-terminal domain of 189 amino acids is predicted to be globular and the central motor domain is also predicted to be largely globular, whereas the C-terminal region is predicted to be largely α-helical (Figs. 2 and 3). KIF2 forms a homodimer and EM revealed KIF2 to be a globular molecule with a diameter of ~16 nm (Fig. 4). KIF2 moves on the microtubule towards the plus end at a velocity of ~0.4 μm/s and accumulates in the proximal region of ligated nerves, thus indicating that KIF2 functions as an anterograde motor in vivo [32]. KIF2 is expressed abundantly in nerve tissues, and expressed ubiquitously in other tissues. The level of expression of KIF2 is regulated developmentally; KIF2 is abundantly expressed in developing axons, whereas the level of expression decreases in adults. Soluble KIF2 does not have a light chain and its cargo, immunoprecipitated by an anti-KIF2 antibody, is vesicles about 100—120 nm in diameter that are distinct from the membranous organelles carried by other motors, such as kinesin and KIF3, and from synaptic vesicle precursors. Because KIF2 is associated with vesicles in growth cone particles, it was suggested that KIF2 is an anterograde motor for the transport of membranous organelles important for neurite extension [32]. Thus, KIF2 is a unique central motor domain type KIF anterogradely transporting vesicles important for axonal extension in developing neurons (Fig. 5).

KIFC2

KIFC2 is a COOH-terminal motor domain type motor for transport of multivesicular body like membrane organelle in the dendrites.

KIFC2 is the first COOH-terminal motor domain type KIF functioning in organelle transport in cells. Several COOH-terminal motor domain type motors such as ncd in *Drosophila* and Kar 3 in *S. cervisiae* were identified and characterized as motors for meiosis, mitosis and karyogamy [18—21]. These C-terminal motor domain type members exhibited microtubule minus end directed motility. Because the only microtubule minus end directed motor identified for organelle transports is cytoplasmic dynein, it is reasonable to assume that as yet unidentified C-terminal KIFs exist for retrograde transport. Based on this assumption searches for new C-terminal motor domain type KIFs have been carried out. Taking into consideration the discovery of a C-terminal KIF consensus sequence at the neck region upstream of the head region [33] and the results of PCR using a motor domain gene consensus sequence [33,34], three C-terminal motor domain type KIFs have been identified in mouse brain [33]. Although KIFC1 (609 aa) is the murine homologue of CHO2 antigen [33], KIFC2 (792 aa) was the first neuron-specific C-terminal motor domain type KIF identified [33,34]. Its motor domain is located at the C-terminal and there is a long α-helical coiled-coil stretch in the stalk (Figs. 2 and 3) [33,34]. KIFC2 forms a homodimer without associating polypeptides. It exhibits a microtubule-activated ATPase activity and is localized, as determined by immunofluorescene analysis, mainly in the cell body and dendrites [33]. A quantitative immunoblotting study revealed that the amount of KIFC2 in the peripheral nerve is very low, only 0.1—0.3% of that in the cerebral cortex [33]. The cargo of KIFC2 was identified as a new kind of multivesicular body-like membranous organelle, which is distinct from conventional multivesicular bodies, functioning as a shuttle between early and late endosomes [33]. Because KIFC2 is mainly localized in the cell body and dendrites, even following the overexpression of KIFC2 in primary cultured neurons, it was strongly suggested that KIFC2 is a unique C-terminal motor domain type motor mainly functioning in the dendritic transport of multivesicular body-like membranous organelles (Fig. 5). Although the amino acid sequence analysis of KIFC2 strongly suggests that it moves along microtubules toward the minus end, it is necessary to carry out an in vitro motility assay to firmly establish this [33,34].

New KIFs

Because we still need to identify the KIFs that play a role in organelle transport such as the KIFs that convey synaptic plasma membrane proteins essential for vesicle docking (e.g., SNAP25, syntaxin 1A), the KIFCs which act as retrograde motors in axons, and the slow transport motors that carry cytoskeletal proteins and cytosolic proteins, the systematic search for cDNA encoding of yet unidentified KIFs has been performed using mouse nerve tissue, kidney, and small intestine cDNA libraries. Using this approach a total of 16 new KIFs (KIF3C, 6, 7, 8, 9, 10, 11, 12, 13A, 13B, 14, 15, 16A, 16B, 17, C3) have been discovered [35]. On the basis of the expression of the mRNA of these KIFs in tissue, these KIFs could play an important role in organelle transport in many types of cells including neurons and epithelial cells.

Conclusion and future challenge

It was determined that a cell uses a surprisingly large number of microtubule associated motor proteins, more than 20 KIFs, which very precisely control the direction and velocity of the transports of various kinds of cargoes. These cargoes include distinct types of membranous organelles, possibly protein complexes and mRNA [1,2,36,37]. The control of transport involves developmental regulation. Because important motors still remain to be identified we need to search for them. The characterization of newly identified members, the analyses of the functions of each member using molecular cell biological and molecular genetic approaches and the understanding of the mechanisms of the recognition of the correct cargoes by each motor are also very important. Furthermore, how and where the cargoes bind to their correct motors and how the cargoes are dissociated from the motors at their destination are also intriguing topics for future research. We are beginning to obtain new information regarding the mechanism of organelle transport in cells, while at the same time new fascinating questions are arising which will lead to very exciting research in this field [38—40].

Acknowledgements

I thank Haruyo Fukuda for her secretarial assistance. This work was supported by a grant for the Center of Excellence (COE) from the Ministry of Education, Science and Culture to N. Hirokawa.

References

1. Hirokawa N. Organelle transport along microtubules — the role of KIFs (Kinesin superfamily proteins). Trends Cell Biol 1996;6:135—141.
2. Hirokawa N. Kinesin and dynein superfamily proteins and the mechanism of organelle transport. Science 1998;279:519—526.
3. Grafstein B, Forman DS. Intracellular transport in neurons. Physiol Rev 1980;60:1167—1283.
4. Hirokawa N. The crosslinker system between neurofilaments, microtubules and membranous organelles in frog axons revealed by quick freeze, freeze fracture, deep etching method. J Cell Biol 1982;94:129—142.
5. Brady ST, Lasek RJ, Allen RD. Fast axonal transport in extruded axoplasm from squid giant axon. Science 1982;218:1129—1131.
6. Vale RD, Reese TS, Sheetz MS. Identification of a novel force-generating protein, Kinesin, involved in microtubule-based motility. Cell 1985;42:39—50.
7. Brady ST. A novel brain ATPase with properties expected for the fast axonal transport motor. Nature 1985;317:73—75.
8. Hirokawa N, Pfister KK, Yorifuji H, Wagner MC, Brady ST, Bloom GS. Submolecular domains of bovine brain kinesin identified by electron microscopy and monoclonal antibody decoration. Cell 1989;56:867—878.
9. Scholey JM, Heuser J, Yang JT, Goldstein LSB. Identification of globular mechanochemical heads of kinesin. Nature 1989;338:355—357.
10. Yang JT, Laymon RA, Goldstein LSB. A three-domain structure of kinesin heavy chain revealed by DNA sequence and microtubule binding analyses. Cell 1989;56:879—889.
11. Pfister KK, Wagner MC, Stenoien DL, Brady ST, Bloom GS. Monoclonal antibodies to kinesin heavy and light chains stain vesicle-like structures, but not microtubules, in cultured cells. J Cell Biol 1989;108:1453—1463.
12. Hirokawa N, Sato-Yoshitake R, Kobayashi N, Pfister KK, Bloom GH, Brady ST. Kinesin associates with anterogradely transported membranous organelles in vivo. J Cell Biol 1991;114: 295—302.
13. Paschal BM, Shpetner HS, Vallee RB. MAP1C is a microtubule-activated ATPase which translocates microtubules in vitro and has dynein-like properties. J Cell Biol 1987;105:1273—1282.
14. Lye M, Porter E, Scholey JM, McIntosh JR. Identification of a microtubule-based cytoplasmic motor in the nematode C. elegans. Cell 1987;51:309—318.
15. Shroer TA, Steuer ER, Sheetz MP. Cytoplasmic dynein is a minus end-directed motor for membranous organelles. Cell 1989;56:937—946.

16. Schnapp BJ, Reese TS. Dynein is the motor for retrograde axonal transport of organelles. Proc Natl Acad Sci USA 1989;86:1548–1552.
17. Hirokawa N, Sato-Yoshitake R, Yoshida T, Kawashima T, Yoshida T. Brain dynein (MAP1C) localizes on both anterogradely and retrogradely transported membranous organelles. J Cell Biol 1990;111:1027–1037.
18. Enos AP, Morris NR. Mutation of a gene that encodes a kinesin-like protein blocks nuclear division in A. nidulans. Cell 1990;60:1019–1027.
19. Meluh PB, Rose MD. KAR3, a kinesin-related gene required for yeast nuclear fusion. Cell 1990; 60:1029–1041.
20. McDonald HB, Goldstein LSB. Identification and characterization of a gene encoding a kinesin-like protein in drosophila. Cell 1990;61:991–1000.
21. Endow SA, Henikiff S, Soler Niedziela L. Mediation of meiotic and early mitotic chromosome segregation in Drosophila by a protein related to kinesin. Nature 1990;345:81–83.
22. Aizawa H, Sekine Y, Takemura R, Zhang Z, Nangaku M, Hirokawa N. Kinesin family in murine central nervous system. J Cell Biol 1992;119:1287–1296.
23. Okada Y, Yamazaki H, Sekine Y, Hirokawa N. The neuron-specific kinesin superfamily protein KIF1A is a unique monomeric motor for anterograde axonal transport of synaptic vesicle precursors. Cell 1995;81:769–780.
24. Hall DH, Hedgecock EM. Kinesin-related gene unc-104 is required for axonal transport of synaptic vesicles in C. elegans. Cell 1991;65:837–847.
25. Otsuka AJ, Jeyaprakash A, Garcia-Anoveros J, Tang LZ, Fisk G, Hartshorne T, Franco R, Born T. The C. elegans unc-104 gene encodes a putative kinesin heavy chain-like protein. Neuron 1991;6:113–122.
26. Yonekawa Y, Harada A, Okada Y, Kanai Y, Takei Y, Funakoshi T, Terada S, Noda T, Hirokawa N. Targeted disruption of the axonal motor protein KIF1A results in impairment of synaptic vesicle precursor transport, motor and sensory neuronal disturbances and neuronal cell death. J Cell Biol 1998;141:431–441.
27. Nangaku M, Sato-Yoshitake R, Okada Y, Noda Y, Takemura R, Yamazaki H, Hirokawa N. KIF1B; A novel microtubule plus end-directed monomeric motor protein for transport of mitochonndria. Cell 1994;79:1209–1220.
28. Kondo S, Sato-Yoshitake R, Noda Y, Aizawa H, Nakata T, Matsuura Y, Hirokawa N. KIF3A is a new microtubules-based anterograde motor in the nerve axon. J Cell Biol 1994;125:1095–1107.
29. Yamazaki H, Nakata T, Okada Y, Hirokawa N. KIF3A/B: A heterodimeric kinesin superfamily protein that works as a microtubule plus end-directed motor for membrane organelle transport. J Cell Biol 1995;130:1387–1399.
30. Yamazaki H, Nakata T, Okada Y, Hirokawa N. Cloning and characterization of KAP3: a novel kinesin superfamily associated protein of KIF3A/3B. Proc Natl Acad Sci USA 1996;93: 8443–8448.
31. Sekine Y, Okada Y, Kondo S, Aizawa H, Takemura R, Hirokawa N. A novel microtubule-based motor protein (KIF4) for organelle transports, whose expression is regulated developmentally. J Cell Biol 1994;127:187–202.
32. Noda Y, Sato-Yoshitake R, Kondo S, Nangaku M, Hirokawa N. KIF2 is a new microtubule-based anterograde motor that transports membranous organelles distinct from those carried by kinesin heavy chain or KIF3A/B. J Cell Biol 1995;129:157–167.
33. Saito N, Okada Y, Noda Y, Kinoshita Y, Kondo S, Hirokawa N. KIFC2 is a novel neuron-specific C-terminal type kinesin superfamily motor for dendritic transport of multivesicular body-like organelles. Neuron 1997;18:425–438.
34. Hanlon DW, Yang Z, Goldstein LSB. Characterization of KIFC2, a neuronal kinesin superfamily member in mouse. Neuron 1997;18:439–451.
35. Nakagawa T, Tanaka Y, Matsuoka E, Kondo S, Okada Y, Noda Y, Kanai Y, Hirokawa N. Identification and classification of sixteen new kinesin superfamily (KIF) proteins in mouse genome. Proc Natl Acad Sci USA 1997;94:9654–9659.

36. Terada S, Nakata T, Peterson A, Hirokawa N. Visualization of slow axonal transport in vivo. Science 1996;273:784–788.
37. Singer RH. RNA: traffic report. Trends Cell Biol 1996;6:486–489.
38. Hirokawa N. Kinesis and dynein superfamily proteins and the mechanism of organelle transport. Science 1998;279:519–526.
39. Harada A, Takei Y, Kanai Y, Tanaka Y, Nonaka S, Hirokawa N. Golgi vesiculation and lysosome dispersion in cells lacking cytoplasmic dynein. J Cell Biol 1998;141:51–59.
40. Tanaka Y, Kanai Y, Okada Y, Nonaka S, Takeda S, Harada A, Hirokawa N. Targeted disruption of mouse conventional kinesin heavy chain, KIF5B, results in abnormal perinuclear clustering of mitochondria. Cell 1998;93:1147–1158.

Dynamic Cells: Cell Biology of the 21st Century
S. Yagihashi, T. Kachi and M. Wakui, editors.

Evolution and regulation of neurosecretory cells

Horst-W. Korf
Dr. Senckenbergische Anatomie, Institut für Anatomie II, Klinikum der Johann Wolfgang Goethe-Universität, Frankfurt/Main, Germany

Abstract. This contribution considers evolutionary and regulatory aspects of neurosecretory cells, using the melatonin-producing pinealocytes as model. Pinealocytes have undergone a striking transformation during phylogeny; they appear as classical and modified photoreceptors in submammalian vertebrates and as pinealocytes sensu stricto in mammals. All types of pinealocyte produce and secrete melatonin in a rhythmic fashion during darkness. This rhythm primarily depends on a single enzyme, the serotonin N-acetyltransferase (NAT), whose rhythmic activation and inactivation is regulated by environmental light stimuli and by signals from endogenous (circadian) oscillator(s). In submammalian species, the light stimuli are directly perceived by the melatonin-producing cells which may also harbor the endogenous oscillator. In mammals the photoreceptors controlling pineal melatonin biosynthesis are located in the retina and the endogenous oscillator resides within the hypothalamic suprachiasmatic nucleus (SCN). Signals from the retina and the SCN are transmitted to the mammalian pineal via the sympathetic innervation whose primary neurotransmitter, norepinephrine, shapes the melatonin rhythm through activation of α_1- and β-adrenergic receptors. In all vertebrate species cyclic AMP and calcium ions have been identified as second messengers that stimulate melatonin biosynthesis, but their regulation shows great variation among the species. Species differences are also evident with regard to the regulation of NAT at the transcriptional and post-transcriptional levels. The results suggest that the receptor mechanisms of pinealocytes are transformed during phylogeny, whereas the neuroendocrine effector mechanism, the nocturnal production of melatonin, is well conserved.

Keywords: adrenoreception, melatonin, pinealocytes, photoreception, transcription factors.

Introduction

Since the pioneering investigations of Ernst and Berta Scharrer [1—4], it is well known that neurosecretory cells possess a high phylogenetic age and are widely distributed in the brain of both invertebrates and vertebrates (Fig. 1). An essential feature of neurosecretory cells is their capacity to produce and dispatch neurohormones into body fluids (hemolymph, blood stream, cerebrospinal fluid) in response to external (environmental) and internal stimuli. Thus, neurosecretory cells form an important bridge between the central nervous system and the endocrine and immune systems in the periphery of the body. This contribution will consider some evolutionary and regulatory aspects of neurosecretory cells using melatonin-producing pinealocytes as model.

Address for correspondence: Horst-W. Korf, Dr. Senckenbergische Anatomie, Anatomisches Institut II, Theodor-Stern-Kai 7, D-60590 Frankfurt, Germany. Tel.: +49-69-6301-6040. Fax: +49-69-6301-6017. E-mail: korf@em.uni-frankfurt.de

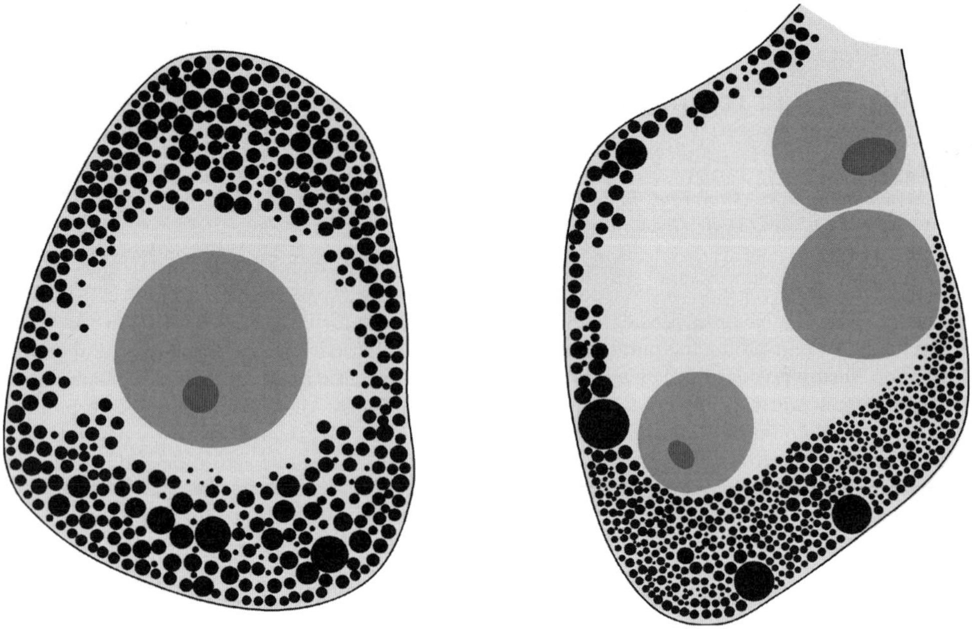

Fig. 1. Schematic drawings of neurosecretory cells in cerebral ganglion of *Aplysia limacina* (left) and the magnocellular preoptic nucleus of the teleost *Centropistes striatus*. After Ernst and Berta Scharrer [1—3].

Evolutionary, structural and functional aspects of pinealocytes

Pinealocytes form the major parenchymal component of the vertebrate pineal organ. These cells have undergone a striking transformation in the course of evolution and appear as classical and modified photoreceptors in submammalian vertebrates and as pinealocytes sensu stricto in mammals (Fig. 2) [5—7]. The classical pineal photoreceptor, which is the predominant type of pinealocyte in poikilothermic vertebrates, is characterized by an outer segment protruding into the pineal lumen and comprising numerous disks. Opposite to the outer segment, the pineal photoreceptor gives rise to a basal process which contains numerous electron lucent synaptic vesicles intermingled with scattered dense core vesicles and synaptic ribbons. Via these processes, the classical pineal photoreceptors establish synapses with intrapineal second-order neurons. Modified pineal photoreceptors primarily found in lacertilian and avian species possess either a rudimentary outer segment or only a bulbous cilium of the 9 + 0 type. Their basal processes which contain synaptic ribbons and numerous dense core vesicles do not form synapses with second-order neurons, but terminate freely at the basal lamina of the pineal organ. Pinealocytes sensu stricto form the parenchyma of the mammalian pineal organ; they lack outer segments, but may contain cilia of the 9 + 0 type as remnants of the receptor pole. The majority of these cells bear

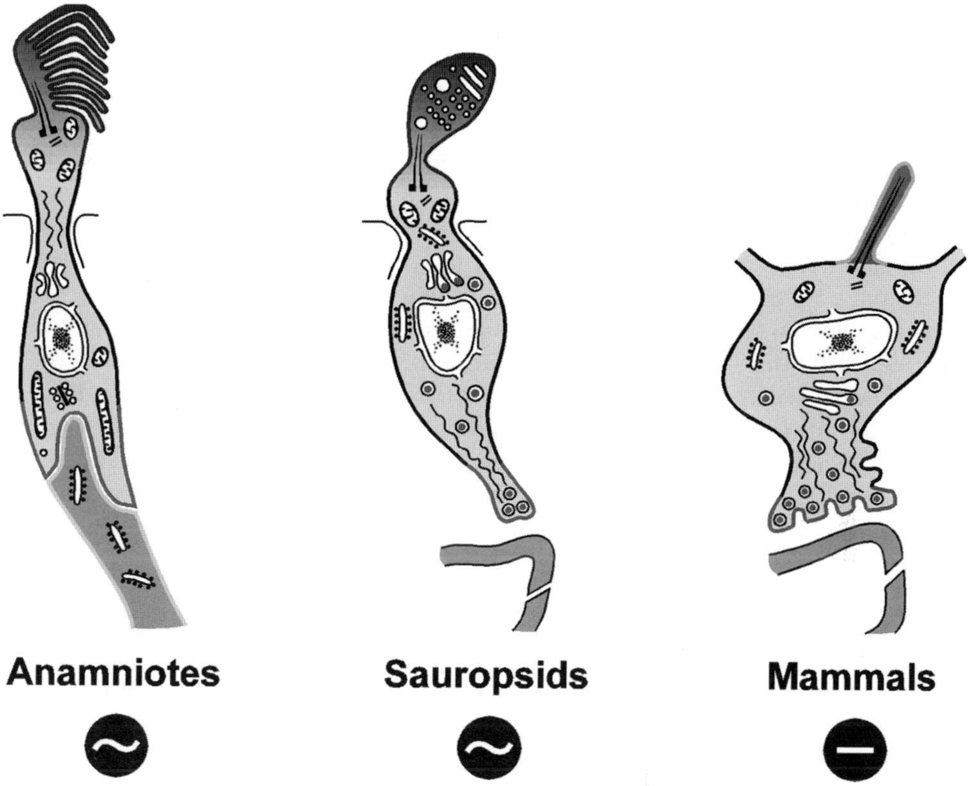

Fig. 2. The three principal types of pinealocytes. Classical pineal photoreceptors of anamniotes; modified photoreceptors of sauropsids and pinealocytes sensu stricto in mammals. Classical and modified pineal photoreceptors may harbor an endogenous oscillator (~) which is absent from mammalian pinealocytes (—).

processes which terminate within the pineal organ at the basal lamina of the perivascular space. These processes contain synaptic ribbons and a varying number of clear and dense core vesicles.

Physiological and biochemical studies have revealed that classical and modified pineal photoreceptors are capable of perceiving and transmitting light stimuli via a phototransduction cascade whose design closely resembles that found in retinal photoreceptors [6,7]. In general, pineal photoreceptors may contain photopigments that occur also in retinal rods or cones, but in some species, e.g., the chicken, most of the pineal photoreceptor cells have been shown to contain a pineal-specific photopigment denominated as pinopsin [8,9].

Mammalian pinealocytes have lost the capacity to perceive and transmit light stimuli, they lack 11-cis retinal, the functional group of any vertebrate photopigment [10,11]. Nevertheless, they express some molecules of the phototransduction cascade, e.g., S-antigen (arrestin), whose functional role, however, remains enigmatic. It has been suggested that these molecules become involved in the transduction of adrenergic signals which are of major importance for the regulation of pineal function in mammals (see below).

Despite the phylogenetic transformation described above, the various types of pinealocytes function as neurosecretory cells which produce and secrete melatonin in all vertebrate species. The melatonin biosynthesis displays a striking rhythm leading to highly elevated melatonin levels during the night and nearly undetectable melatonin levels during the day. Melatonin can thus be considered as the neuroendocrine messenger of darkness and a timing hormone. According to current concepts, the highly lipophilic melatonin is not stored within the pinealocytes, but is released into the circulation immediately upon its formation. Consequently, the secretion of melatonin appears to be controlled solely by its biosynthesis which involves identical enzymes in all vertebrate species (Fig. 3). The rate-limiting enzyme is the serotonin-N-acetyltransferase (NAT) which appears as a molecular target for all regulatory inputs. This enzyme is activated every night by means of transcriptional, translational and posttranslational mechanisms and thus causes the nocturnal surge in melatonin production and secretion.

The NAT rhythm is regulated by the ambient lighting conditions (the photoperiod) and signals from an endogenous oscillator which generates a "circadian" rhythm that is independent from any environmental cue (Zeitgeber) and whose period is approximately the length of a day. In submammalian vertebrates, environmental light stimuli can be directly perceived by the melatonin-producing cells, i.e., the classical and modified pineal photoreceptors. As shown for several

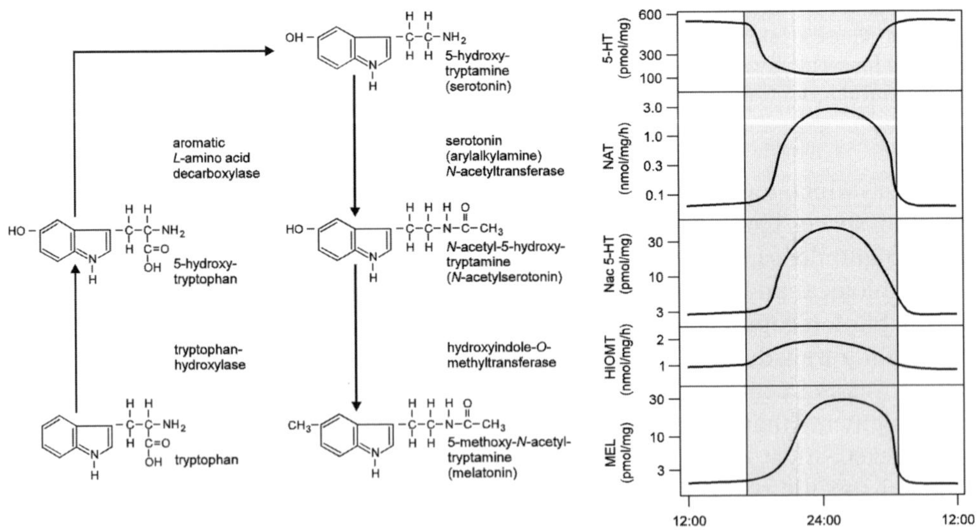

Fig. 3. The melatonin biosynthesis and its enzymes are shown on the left. The right panel depicts the diurnal rhythms in the concentration of serotonin (5-HT), N-acetylserotonin (Nac 5-HT), melatonin (MEL) and in the activity of the serotonin-N-acetyltransferase (NAT) and the hydroxyindole-*O*-methyltransferase (HIOMT) as determined in the rat pineal organ. (After Klein 1985, from [7]).

submammalian species (e.g., lamprey, zebra fish, pike, sparrow, chicken) the pineal organ also harbors an endogenous oscillator, and it is conceivable that, in these animals, photoreception, circadian rhythm generation and melatonin production are accomplished within a single cell.

In birds, the pineal organ receives a dense sympathetic innervation and norepinephrine (NE), the primary neurotransmitter of this innervation appears to modulate the inherent rhythmicity of the pineal organ. In contrast to mammals (see below), NE is released from the sympathetic nerve terminals in the avian pineal organ during daytime and it inhibits the pineal melatonin biosynthesis via activation of α_2-adrenergic receptors.

In mammals, the pineal organ possesses neither photoreceptors nor endogenous rhythm generators. The photoreceptor cells controlling melatonin biosynthesis are located in the retina, but it remains open whether these cells are rods or cones or belong to another type of photoreceptor which has not yet been identified. The endogenous oscillator is located in the bilaterally arranged suprachiasmatic nuclei (SCN) of the hypothalamus, where single neurons are capable of generating a circadian rhythm [12]. This circadian rhythm is synchronized with the photoperiod through signals which are transmitted from the retina to the SCN via the retinohypothalamic tract. The SCN conveys its information to mammalian pineal organ via a complex neuronal circuit whose final pathway is the sympathetic innervation of the mammalian pineal organ, which originates from the superior cervial ganglia (Fig. 4). In contrast to all other vertebrate classes, this sympathetic innervation is mandatory for generating and maintaining the rhythm of the pineal melatonin biosynthesis in mammals. Its primary neurotransmitter, norepinephrine, is rhythmically released in large quantities at the onset of darkness and stimulates NAT and melatonin biosynthesis through activation of α_1- and β-adrenergic receptors [13].

Transmembrane signaling cascades and second messengers regulating melatonin biosynthesis

As shown by numerous comparative investigations, the pineal melatonin biosynthesis of poikilothermic vertebrates is primarily regulated through photoreception and phototransduction, whereas, in mammals, the melatonin biosynthesis is controlled through reception and transduction of noradrenergic signals. The avian pineal represents an intermediate form that employs both photoreception/-transduction and adrenoreception/-transduction as regulatory processes. In all vertebrate species investigated thus far, cyclic AMP and calcium ions have been identified as the second messengers that stimulate NAT and melatonin biosynthesis, but as can be expected from the evolutionary transformations, the regulation of these second messengers differs from species to species. In the following, these differences will be pointed out using three different vertebrate species: the rainbow trout (whose pineal contains classical photoreceptors and lacks the sympathetic innervation), the chicken (whose pineal contains modified pineal

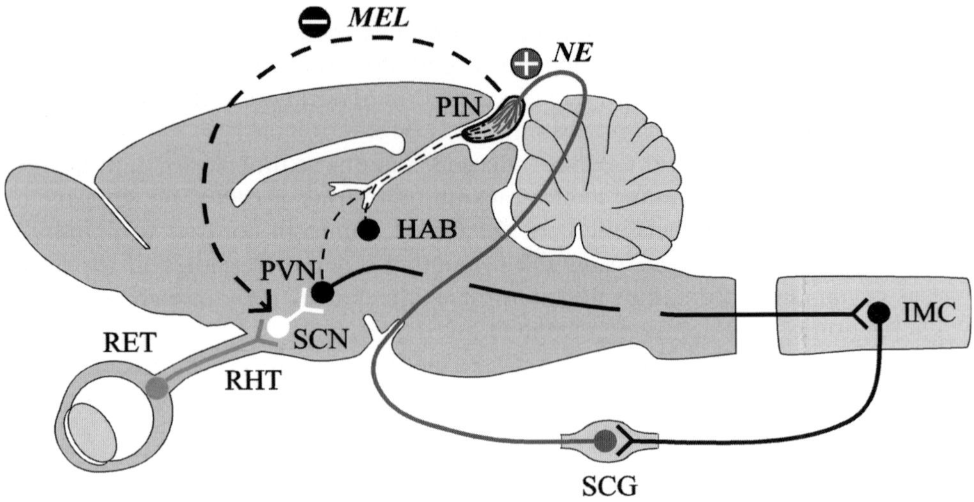

Fig. 4. The melatonin rhythm generating system of the rat. Ganglion cells of the retina (RET) innervate the hypothalamic suprachiasmatic nucleus (SCN) via the retinohypothalamic tract (RHT). The SCN harbors the circadian oscillator. It projects to the paraventricular nucleus (PVN) that gives rise to two pinealopetal pathways: 1) The central innervation (dashed lines) reaches the pineal gland directly via its stalk; fibers from the habenular nucleus (HAB) join this projection. 2) The PVN forms a monosynaptic connection to the intermediolateral column (IMC) in the thoracic spinal cord, which gives rise to preganglionic sympathetic nerve fibers projecting to the superior cervical ganglion (SCG). The SCG is the origin of postganglionic nerve fibers which innervate the pineal organ (PIN) and release the neurotransmitter norepinephrine (NE) in a circadian fashion. NE activates the synthesis of melatonin (MEL) which feeds back to inhibit SCN activity. (From [7]).

photoreceptors and receives a sympathetic innervation acting upon α_2-adrenergic receptors), and the rat (whose pineal lacks photoreceptors and receives a sympathetic innervation acting upon α_1- and β-adrenergic receptors).

Rainbow trout

Experiments with cultured trout pineal organs have shown that the melatonin secretion follows a robust, irradiance-dependent secretion profile. Attenuation of light irradiance is associated with a distinct increase in melatonin production whereas enhancement of the illumination is followed by a clear decrease [14]. Elevation of the intracellular cAMP concentration was demonstrated to stimulate NAT activity and melatonin production [15]. Calcium ions appear as important mediators of the darkness-induced increase in melatonin production: elevation of the intracellular calcium ion concentration ($[Ca^{2+}]_i$) stimulated melatonin biosynthesis whereas blocking of calcium channels by nifedipine, verapamil or cobalt chloride abolished the increase in Ca_i^{2+} probably by action of these substances on voltage-gated L-type calcium channels [14,16,17]. Calcium imaging of isolated trout pinealocytes identified by the S-antigen immunoreaction revealed that 30% of the pinealocytes exhibited spontaneous $[Ca^{2+}]_i$ oscillations whose frequency differed from cell to cell (Fig. 5). These oscillations were independent of the lighting conditions and were not evoked by light-induced activation of phototransduction processes. The trigger for the $[Ca^{2+}]_i$ oscillations remains unknown. Since removal of extracellular calcium or application of nifedipine caused a reversible breakdown of these oscillations, they depend on the influx of extracellular calcium ions through dihydropyridine-sensitive voltage-

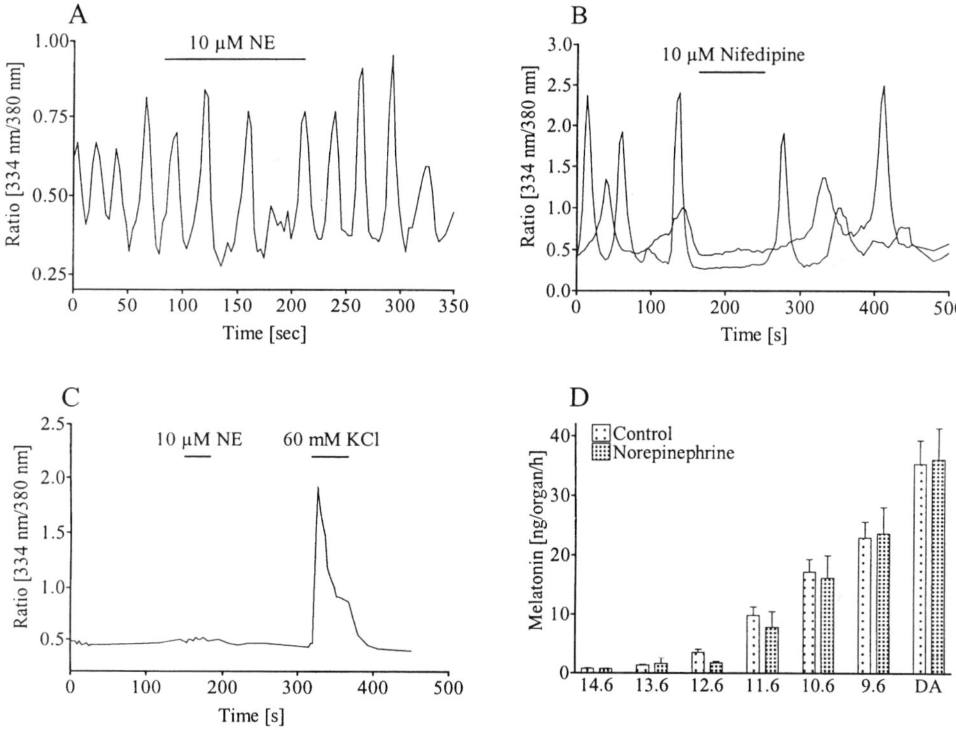

Fig. 5. Regulation of the intracellular calcium ion concentration ($[Ca^{2+}]_i$) in S-antigen immunoreactive trout pinealocytes and control of melatonin biosynthesis in the trout pineal organ. **A,B:** Several S-antigen immunoreactive trout pinealocytes (up to 30% of the total population) displayed spontaneous $[Ca^{2+}]_i$ oscillations of varying frequencies. **A:** These oscillations were not affected by treatment of the cells with 10 μM norepinephrine. **B:** Inhibition of voltage-gated calcium channels of the L-type with 10 μM nifedipine blocked the spontaneous $[Ca^{2+}]_i$ oscillations. After removal of the drug, the $[Ca^{2+}]_i$ oscillations started again. **C:** The majority of the S-antigen-immunoreactive trout pinealocytes did not show spontaneous $[Ca^{2+}]_i$ oscillations but had stable basal $[Ca^{2+}]_i$ levels. Also in these cells whose excitability was proven by depolarizing them with 60 mM KCl norepinephrine did not influence the $[Ca^{2+}]_i$ levels. **A–C:** adapted from [17]; duration of the drug treatment is indicated by bars. **D:** Control of melatonin release from isolated superfused trout pineal organs. The release of melatonin depends on the illumination (light intensity varying from 14.6 to 9.6 log photons/cm^2/s) and is maximal in dark-adapted (DA) organs. Norepinephrine does not influence melatonin secretion. Mean ± SEM of n = 8 (controls) or n = 4 (norepinephrine). (Courtesy of H. Meissl, Bad Nauheim; cf. [14], from [7]).

gated calcium channels that are obviously opened and closed in a rhythmic manner. The majority of the S-antigen-immunoreactive trout pinealocytes did not display spontaneous $[Ca^{2+}]_i$ oscillations, but also these cells were endowed with voltage-gated calcium channels of the L-type. Taken together, all data underline an important role of calcium ions and voltage-gated calcium channels of the L-type for regulation of melatonin biosynthesis in the trout. The elevation of $[Ca^{2+}]_i$ may influence melatonin biosynthesis through its interaction with the cAMP pathway, but also through other mechanisms.

Norepinephrine affected neither the maximal release of melatonin in darkness nor the melatonin response in the photopic and mesopic range of illumination [14]. Also the intracellular concentration of cyclic AMP and calcium ions were not affected by norepinephrine [15,17].

Chicken

The melatonin biosynthesis in the chicken pineal organ is primarily regulated by photoreceptors and endogenous oscillators located in the pineal organ itself. Chicken pineal organs kept in vitro and deprived of any neuronal input are able to maintain an endogenous rhythm in melatonin production [18,19]. Discrete light pulses cause an acute inhibition of the melatonin biosynthesis and entrain the endogenous oscillator. The acute inhibition of melatonin production appears to be mediated by cAMP, whereas the entraining effect of light is not [20,21]. cAMP levels oscillate with high values during the scotophase and low values during the day.

Calcium ions play an important role for regulation of melatonin biosynthesis in chicken, since the nocturnal increase in melatonin production is suppressed by calcium channel blockers [22—24]. Calcium influx appears to regulate cAMP levels, suggesting that calcium and cAMP interact to regulate the activity of NAT.

Calcium imaging analysis of isolated chicken pinealocytes [25] revealed that approximately 10% of the pinealocytes exhibited spontaneous $[Ca^{2+}]_i$ oscillations. The pools required for the maintenance of these oscillations have not yet been identified, but it may well be that, like in the trout, voltage-gated calcium channels are involved. NE had no effect on $[Ca^{2+}]_i$ in chicken pinealocytes. Agents that increase intracellular cyclic nucleotides cause increase in $[Ca^{2+}]_i$ levels probably via activation of cyclic nucleotide-gated channels of the rod- or cone-type [26,27].

Using the patch clamp technique D'Souza and Dryer [28] have discovered a novel cationic channel in chicken pinealocytes which has an unusually long opening time and is therefore denominated as I_{LOT}. This channel is nonselective for cations and does not appear to be gated by voltage or soluble second messengers. The channel is active at night, but not during the day. Interestingly, the rhythm in channel activity persists in pinealocytes kept in constant darkness and is thus under the control of the circadian oscillator. Because the channel is nonselective for cations, chicken pinealocytes would be more depolarized at night and calcium ions could enter the cells through both voltage-sensitive channels and I_{LOT} itself. The elevation of $[Ca^{2+}]_i$ would then cause an elevation of intracellular cAMP via activation of calcium/calmodulin stimulated adenylate cyclase and the joint increase in calcium ions and cAMP would optimally stimulate melatonin biosynthesis at night [29].

The sympathetic innervation and its primary neurotransmitter norepinephrine appear to modulate the inherent rhythmicity of the chicken pineal organ. In contrast to mammals (see below) NE inhibits the pineal melatonin biosynthesis in chicken [23]. The opposite effects of NE in birds and mammals can be readily explained by the fact that the receptors activated by NE differ in the chicken and mammalian pineal. Whereas in mammals NE acts upon α_1- and β-receptors, the chicken pineal organ contains α_2-adrenoreceptors which are coupled to the GTP-binding protein G_i; thus, their activation causes an inhibition of the adenylate cyclase activity and a drop in intracellular cAMP, finally inhibiting the melatonin biosynthesis.

Rat

Like in all mammals, NE appears as the essential regulator of the melatonin biosynthesis in rat. NE, released in large quantities from the sympathetic nerve terminals after the onset of darkness acts upon α_1- and β-adrenergic receptors in the pinealocyte membrane (Fig. 6) [13]. Activation of β-receptors is an absolute requirement for stimulation of melatonin biosynthesis. This leads to an increase in adenylate cyclase activity resulting in a 10-fold increase in the intracellular concentration of cAMP which is the essential second messenger required for the nocturnal rise in melatonin production. Activation of the β-receptor is mediated via the GTP-binding protein G_s whose α-subunit has been found in the rat pineal in relatively high concentrations.

Stimulation of α_1-adrenergic receptors has no effect on cAMP accumulation and melatonin biosynthesis on its own, but potentiates the β-adrenergic effect: simultaneous activation of α_1- and β-adrenergic receptors causes a 100-fold increase in cAMP and a maximal induction of the melatonin biosynthesis. The potentiation is based on a rise in $[Ca^{2+}]_i$ induced by α_1-adrenergic stimulation. Calcium imaging has shown that activation of α_1-adrenergic receptors evokes a biphasic rise in $[Ca^{2+}]_i$ in more than 95% of the pinealocytes which were identified by means of the S-antigen immunoreaction. The primary phase reflects mobilization of calcium ions from intracellular stores whereas the secondary phase involves an increase in calcium influx. The channel responsible for this influx has not yet been identified. Voltage-gated calcium channels of the L-type are present in rat pinealocytes, but they are apparently not involved in the NE-induced calcium signaling, since their blockage by nifedipine did not affect the NE-induced calcium signal [30,31].

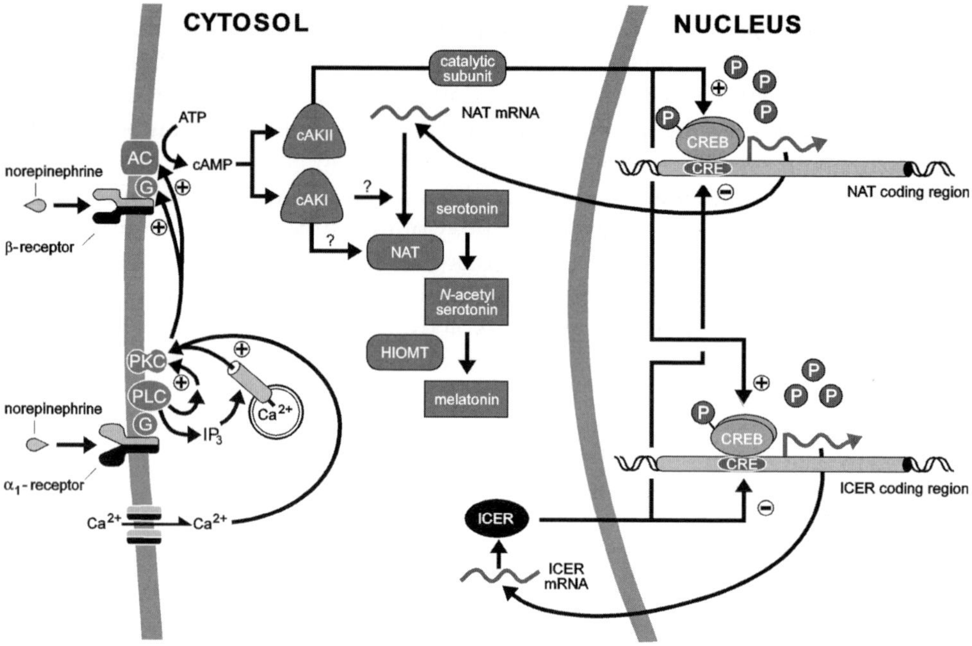

Fig. 6. Schematic diagram of the noradrenergic signal transduction cascade and its second and third messengers in rat pinealocytes. AC: adenylate cyclase. G: GTP-binding protein. PKC: protein kinase C. PLC: phospholipase C. IP_3: inositoltrisphosphate. cAKI, I-I cyclic AMP-dependent protein kinase type I and II. NAT: serotonin (arylalkylamine) N-acetyltransferase. HIOMT: hydroxyindole-O-methyltransferase. CRE: cyclic AMP response element. CREB: CRE binding protein. ICER: inducible cyclic AMP early repressor. P: phosphate groups. (Modified and extended after [44], from [7]).

Activation of NAT expression

Increase and maintenance in NAT activity depend on transcriptional, translational and translational mechanisms regulated by elevated cAMP levels [32]. The importance of transcriptional events for induction of NAT activity varies remarkably on a species-to-species basis. In the sheep the level of NAT mRNA varies between day and night by a factor of 1.5, but the enzyme activity is increased 7-fold in the night [33], indicating the importance of posttranscriptional control mechanisms in this species. By contrast, the 100-fold nocturnal increase of NAT activity in the rat pineal organ is directed transcriptionally by a more than 150-fold increase in NAT mRNA levels at night time [34]. The cAMP-induced transcription of the rat NAT gene rests upon a cyclic AMP response element (CRE) present in the NAT promoter [35].

This CRE appears as a target for the transcription factors CREB (CRE binding protein) and ICER (inducible cAMP early repressor). CREB is a 43-kDa protein that is ubiquitously expressed and is activated by phosphorylation of the amino acid serine at position 133 [36]. Phosphorylated CREB is known to stimulate transcription of genes with CREs in their promoter regions [37]. CREB is expressed in the mammalian pineal gland [38,39] and the cAMP-dependent phosphorylation of CREB has been shown in the rat pineal by gel mobility shift assays, immunoblotting and immunocytochemistry [40,41]. Immunocytochemical demonstration of phosphorylated (p-) CREB in isolated rat pinealocytes revealed that stimulation with norepinephrine or the β-adrenergic agonist isoproterenol results in a similar, time-dependent induction of p-CREB immunoreactivity exclusively located in cell nuclei. The response is observed in more than 95% of the pinealocytes identified by the S-antigen immunoreaction. It reaches its maximum after 30—60 min and remains for up to 300 min, provided the cells are constantly exposed to NE. Upon removal of NE, the p-CREB immunoreaction rapidly declines to background levels within 60 min. Stimulation of α_1-adrenergic receptors does not cause p-CREB induction. Thus, NE-induced phosphorylation of CREB in the rat pineal seems to depend on an activation of the cAMP-protein kinase A pathway. Experiments with agonists and antagonists that allow to distinguish between type I and type II of protein kinase A showed that the phosphorylation of CREB is mediated by protein kinase A type II [42]. These studies also revealed a tight correlation between the activation of protein kinase A type II, the extent of CREB phosphorylation and melatonin secretion. All data suggest that expression of the NAT in the rat pineal organ is pivotally controlled by the NE/cAMP/PKA pathway centered around the phosphorylation of CREB (Fig. 6).

A day/night rhythm in CREB phosphorylation has also been observed in the rat pineal under in vivo conditions. P-CREB levels stay elevated throughout more than half of the night, indicating the need of a maintained high pool of the activated transcription factor for nocturnally elevated pineal metabolism. In the rat, NAT transcription also stays elevated throughout the night and a causal relationship between CRE-mediated transcriptional activation of NAT and the amount of CREB seems likely (E. Maronde, M. Pfeffer, C. Molina, H.-W. Korf and J. Stehle, unpublished results).

Interestingly, the levels of p-CREB did not show light/dark differences in the trout pineal organ (S. Kroeber, E. Maronde and H.-W. Korf, unpublished results). These results provide further evidence for the assumption that considerable species differences exist with regard to the control of NAT expression.

Downregulation of NAT expression

The mechanisms involved in downregulation of NAT transcription occurring at the end of the night are in the focus of intense current research. The level of p-CREB starts to decrease several hours before the end of the night (E. Maronde, M. Pfeffer, C. Molina, H.-W. Korf and J. Stehle, unpublished results). Thus, the potency of this transcription factor to activate gene transcription may fade because of dephosphorylation of CREB. This event is flanked by another transcriptional mechanism, the active downregulation of cAMP-inducible genes by the transcription factor ICER (inducible cAMP early repressor) which functions as a potent inhibitor of the transcription of cAMP-inducible genes [38,39]. ICER is generated as a splicing product of the CREM gene by use of an internal promoter which is located downstream to the promoter which generates all other CREM isoforms known so far. Due to the use of the ICER-specific internal promoter, this CREM transcript is a remarkably small transcription factor comprising the DNA binding domain only. The ICER transcript lacks the amino terminal phosphorylation domain (P-box), although the hallmark of the CREM gene is preserved, i.e., the capacity to generate modular structured isoforms through differential splicing [38]. The cAMP-inducibility of ICER rests upon the presence of four CREs in the 5′ untranslated part of the transcript [43]. ICER shows a superior affinity to CREs and is a far more potent inhibitor of cAMP-inducible genes than its CREM siblings (CREM α, β, γ), when acting as a homodimer. On the other hand, ICER is able to heterodimerize with other CREM isoforms and CREB. By heterodimerization, ICER may impose its superior affinity to CREs on other transcription factors, but it may also scavenge potentially activating CRE-binding proteins. Thus, ICER can regulate target-gene expression in a dualistic way by either indirectly inhibiting gene activation or by directly inhibiting gene transcription. The omission of the phosphorylation domain from ICER indicates that posttranslational modifications do not guide its function and the activity of ICER is solely determined by the level of its expression.

Investigations employing in situ hybridization, Northern blot analysis and RNAse protection analysis revealed a remarkable rhythm in ICER mRNA concentration in the rat pineal gland which is strongly elevated during the night and reaches peak values 7—9 h after onset of darkness. This rhythm persists in constant darkness. Pharmacological in vivo and in vitro experiments clearly showed that the ICER induction strictly depends on the activation of β-adrenergic receptors and the cAMP pathway as holds true for the phosphorylation of CREB and the induction of the melatonin biosynthesis. Moreover, the time course of ICER induction and the presence of four CREs on its promoter region indicate that the ICER transcription is regulated through the PKA-dependent phosphorylation of CREB. Due to the molecular organization of the ICER promoter this CREM isoform can inhibit its own transcription. This autoregulatory feedback loop can be observed in vitro and in vivo and reflects a transient blockade of ICER mRNA accumulation at a time when ICER protein is elevated [38]. Time course analyses of the ICER expression and the NAT activity under natural or in vitro conditions strongly argue in favor of a functional interaction between ICER and NAT [39]. Also the coincident onset of: 1) the daily and circadian rhythms in ICER expression, 2) the ICER inducibility by activation of the cAMP signaling pathway, and 3) the melatonin biosynthesis in the rat pineal gland during ontogenetic development [39] strengthens the view that ICER is required for the downregulation of the NAT at the end of the night [44]. Transfection of rat pinealocytes with an ICER antisense probe revealed that this specific ICER "knockout" resulted in a dramatic increase in NAT mRNA levels and a disinhibited melatonin production (E. Maronde, M. Pfeffer, C. Molina, H.-W. Korf and J. Stehle, unpublished results). All these results make it very probable that, in the rat, ICER plays an important role in the transcriptional downregulation of NAT by its direct action on the promoter of the NAT gene (Fig. 6).

A novel mechanisms of downregulation of NAT activity in the rat pineal gland has been recently discovered by Gastel et al. [45]. These authors demonstrated that the acute inhibition of NAT activity observed after exposure of the rats to light stimuli in the middle of the night does not involve transcriptional and translational control mechanisms, but is mediated by rapid and reversible control of selective proteasomal proteolysis. It will be interesting to determine in further studies whether this mechanism also functions as a molecular switch at the end of the night in animals kept under natural photoperiodic conditions.

Another important issue will be the comparative analysis of mechanisms regulating the expression and activity of NAT in submammalian species whose pineal is directly photosensitive.

Conclusion

In this contribution some evolutionary and regulatory aspects of neurosecretory cells are reviewed, using the melatonin-producing pinealocytes as a model. The results suggest that the neuroendocrine effector mechanism, i.e., the nocturnal production and secretion of the neurohormone melatonin is well conserved in the course of evolution. In contrast, the receptor mechanisms and signal transduction cascades regulating production and secretion of the neurohormone display conspicuous species-to-species differences. This scenario may also hold true for other neurosecretory cell types in both the central and peripheral nervous systems.

Acknowledgements

Experimental studies in the author's laboratory have been supported by the Deutsche Forschungsgemeinschaft.

References

1. Scharrer E. Die Lichtempfindlichkeit blinder Elritzen. Untersuchungen über das Zwischenhirn der Fische. Z vergl Physiol 1928;7:1–38.
2. Scharrer E, Scharrer B. Über Drüsen-Nervenzellen und neurosekretorische Organe bei Wirbellosen und Wirbeltieren. Biol Rev 1937;12:185–216.
3. Scharrer E, Scharrer B. Neuroendocrinology. New York: Columbia Univ Press, 1963.
4. Korf HW, Usadel KH. Neuroendocrinology: Retrospect and Perspectives. Berlin: Springer, 1997.
5. Oksche A, Korf HW, Rodríguez EM. Pinealocytes as photoneuroendocrine units of neuronal origin: concepts and evidence. In: Fraschini F, Reiter RJ (eds) Advances in Pineal Research, vol. 2. London: John Libbey, 1987;1–18.

6. Korf HW. The pineal organ as a component of the biological clock. Ann NY Acad Sci 1994;719:13–42.
7. Korf HW, Schomerus C, Stehle J. The pineal organ, its hormone melatonin and the photoneuroendocrine system. Adv Anat Embryol Cell Biol 1998;146:1–100.
8. Okano T, Yoshizawa T, Fukada Y. Pinopsin is a chicken pineal photoreceptive molecule. Nature 1994;372:94–96.
9. Max M, McKinnon PJ, Seidenman KJ, Barrett RK, Applebury ML, Takahashi JS, Margolskee RF. Pineal opsin: a nonvisual opsin expressed in chick pineal. Science 1995;267:1502–1506.
10. Foster RG, Timmers AM, Schalken JJ, de Grip WJ. A comparison of some photoreceptor characteristics in the pineal and retina. II. The Djungarian hamster (*Phodopus sungorus*). J Comput Physiol 1989;165:565–572.
11. Kramm CM, de Grip WJ, Korf HW. Rod-opsin immunoreaction in the pineal organ of the pigmented mouse does not indicate the presence of a functional photopigment. Cell Tissue Res 1993;274:71–78.
12. Welsh DK, Logothetis DE, Meister M, Reppert SM. Individual neurons dissociated from rat suprachiasmatic nucleus express independently phased circadian firing rhythms. Neuron 1995;14:697–706.
13. Klein DC. Photoneural regulation of the mammalian pineal gland. In: Evered D, Clark S (eds) Photoperiodism, Melatonin and the Pineal Gland. London: Pitman, 1985;38–56.
14. Meissl H, Kroeber S, Yáñez J, Korf HW. Regulation of melatonin production and intracellular calcium concentrations in the trout pineal organ. Cell Tissue Res 1996;286:315–323.
15. Thibault C, Falcón J, Greenhouse SS, Lowery A, Gern WA, Collin JP. Regulation of melatonin production by pineal photoreceptor cells: role of cyclic nucleotides in the trout (*Oncorhynchus mykiss*). J Neurochem 1993;61:332–339.
16. Bégay V, Bois P, Collin JP, Lenfant J, Falcon J. Calcium and melatonin production in dissociated trout pineal photoreceptor cells in culture. Cell Calcium 1994;16:37–46.
17. Kroeber S, Schomerus C, Korf HW. Calcium oscillations in a subpopulation of S-antigen-immunoreactive pinealocytes of the rainbow trout (*Oncorhynchus mykiss*). Brain Res 1997; 744:68–76.
18. Zatz M, Mullen DA, Moskal JR. Photoendocrine transduction in cultured chick pineal cells: effects of light, dark, and potassium on the melatonin rhythm. Brain Res 1988; 438: 199–215.
19. Robertson LM, Takahashi JS. Circadian clock in cell culture: I. Oscillation of melatonin release from dissociated chick pineal cells in flow-through microcarrier culture. J Neurol Sci 1988;8: 2–21.
20. Zatz M, Mullen DA. Norepinephrine, acting via adenylate cyclase, inhibits melatonin output but does not phase-shift the pacemaker in cultured chick pineal cell. Brain Res 1988;450: 137–143.
21. Zatz M, Mullen D. Photoendocrine transduction in cultured chick pineal cells. II. Effects of forskolin, 8-bromocyclic AMP, and 8-bromocyclic GMP on the melatonin rhythm. Brain Res 1988;453:51–62.
22. Harrison N, Zatz M. Voltage-dependent calcium channels regulate melatonin output from cultured chick pineal cells. J Neurol Sci 1989;9:2462–2467.
23. Takahashi JS, Murakami N, Nikaido S, Pratt B, Robertson LM. The avian pineal — a vertebrate model system of the circadian oscillator: cellular regulation of circadian rhythms by light, secondary messengers, and macromolecular synthesis. Recent Prog Horm Res 1989;45:279–352.
24. Zatz M. Agents that affect calcium influx can change cyclic nucleotide levels in cultured chick pineal cells. Brain Res 1992;583:304–307.
25. D'Souza T, Dryer SE. Intracellular free Ca^{2+} in dissociated cells of the chick pineal gland: regulation by membrane depolarization, second messengers and neuromodulators, and evidence for release of intracellular Ca^{2+} stores. Brain Res 1994;656:85–94.
26. Dryer SE, Henderson D. A cyclic GMP-activated channel in dissociated cells of the chick pineal gland. Nature 1991;353:756–758.

27. Bönigk W, Müller F, Middendorff R, Weyand I, Kaupp UB. Two alternatively spliced forms of the cGMP-gated channel a-subunit from the cone photoreceptor are expressed in the chick pineal organ. J Neurosci 1996;16:7458–7468.
28. D'Souza T, Dryer SE. A cationic channel regulated by a vertebrate intrinsic circadian oscillator. Nature 1996; 382: 165–167.
29. Takahashi JS. Ion channels get the message. Nature 1996;382:117–118.
30. Schomerus C, Laedtke E, Korf HW. Calcium responses of isolated, immunocytochemically identified rat pinealocytes to noradrenergic, cholinergic and vasopressinergic stimulations. Neurochem Int 1995;27:163–175.
31. Korf HW, Schomerus C, Maronde E, Stehle JH. Signal transduction molecules in the rat pineal organ: Ca^{2+}, pCREB, and ICER. Naturwissenschaften 1996;83:535–543.
32. Klein DC, Roseboom PH, Coon SL. New light is shining on the melatonin rhythm enzyme. The first postcloning review. Trends Endocrinol Metab 1996;7:106–112.
33. Coon SL, Roseboom PH, Baler R, Weller JL, Namboodiri MAA, Koonin EV, Klein DC. Pineal serotonin-N-acetyltransferase: expression cloning and molecular analysis. Science 1995;270: 1681–1683.
34. Borjigin J, Wang MM, Snyder SH. Diurnal variation in mRNA encoding serotonin-N-acetyltransferase in pineal gland. Nature 1995;378:783–785.
35. Baler R, Covington S, Klein DC. The rat arylalkylamine *N*-acetyltransferase gene promoter. J Biol Chem 1997;272:6979–6985.
36. Gonzales GA, Montminy M. Cyclic AMP stimulates somatostatin gene transcription by phosphorylation of CREB at serine 133. Cell 1989;59:675–680.
37. De Groot RP, Sassone-Corsi P. Hormonal control of gene expression: multiplicity and versatility of cyclic adenosine 3′,5-monophosphate-responsive nuclear regulators. Molec Endocrinol 1993;10:145–153.
38. Stehle JH, Foulkes NS, Molina CA, Simonneaux V, Pévet P, Sassone-Corsi P. Adrenergic signals direct rhythmic expression of transcriptional repressor CREM in the pineal gland. Nature 1993;356:314–320.
39. Stehle JH, Foulkes NS, Pévet P, Sassone-Corsi P. Developmental maturation of pineal gland function: synchronized CREM inducibility and adrenergic stimulation. Molec Endocrinol 1995;9:706–716.
40. Roseboom PH, Klein DC. Norepinephrine stimulation of pineal cyclic AMP response element-binding protein phosphorylation: involvement of a β-adrenergic/cyclic AMP mechanism. Molec Pharmacol 1995;47:439–449.
41. Tamotsu S, Schomerus C, Stehle JH, Roseboom PH, Korf HW. Norepinephrine-induced phosphorylation of the transcription factor CREB in isolated rat pinealocytes: an immunocytochemical study. Cell Tissue Res 1995;282:219–226.
42. Maronde E, Schomerus C, Stehle J, Korf HW. Control of CREB phosphorylation and its role for induction of melatonin synthesis in rat. Biol Cell 1997;89:505–511.
43. Molina C, Foulkes NS, Lalli E, Sassone-Corsi P. Inducibility and negative autoregulation of CREM: an alternative promoter directs the expression of ICER, an early response repressor. Cell 1993;75:1–20.
44. Stehle JH. Pineal gene expression: dawn in a dark matter. J Pineal Res 1995;18:179–190.
45. Gastel JA, Roseboom P, Rinaldi PA, Weller JL, Klein DC. Melatonin production: proteasomal proteolysis in serotonin N-acetyltransferase regulation. Science 1998;279:1358–1360.

Dynamic Cells: Cell Biology of the 21st Century
S. Yagihashi, T. Kachi and M. Wakui, editors.

Dynamic and versatile structures of adrenal medulla, related to pineal and surgery

T. Kachi, G. Takahashi, T. Suzuki, N. Kimura, S. Kajihara, M. Kurushima, T. Yamauchi and T. Irie
Department of Anatomy, Hirosaki University School of Medicine, Hirosaki, Japan

Abstract. Microscopical structures of the adrenal medulla under normal and experimental conditions were reviewed mainly from our work and related areas. The main issues were as follows: 1. Marked differences in the distribution pattern of adrenaline (A) cells and noradrenaline (NA) cells among various mammals were shown and discussed especially in relation to the adaptation and the poor development of A cells and pineal gland in whales. 2. The neuroendocrine complex involving supporting cells in A-cell regions was markedly different from that in NA-cell regions from various points of view. The pineal effect was more apparent in A-cell regions. 3. Various types of nerve-ending plasticity including the invagination complex were shown. 4. Morphological changes at different time points of day following pinealectomy, sham-pinealectomy and melatonin suggested the pineal's phase-tuning effect and the pineal-dependent sham-pinealectomy effect. In addition, observed quantitative changes, both positive and negative, following pinealectomy, were interpreted by means of the intracellular compact cooperation hypothesis. By this mechanism, A cells were postulated to produce epinephrine and peptides such as methionine-enkephalin in a compact and economical manner and enable body cells to adapt to different situations. 5. Direct and indirect mechanisms of pineal actions on the adrenal medulla, including the Ca^{2+}-related mechanisms, have been discussed.

Keywords: calcium, cell-type difference, enkephalin, intracellular cooperation, nerve-ending plasticity, sham-pinealectomy effect.

Introduction

The adrenal medulla and the pineal gland have embryological and biochemical similarities [1,2]. That is, both organs are amine-secreting endocrine glands, having a neuroectodermal origin. Their hormones, epinephrine (E) and melatonin, are finally synthesized by methyltransferase enzymes, PNMT and HIOMT. In addition, the adrenal medulla secretes various peptide hormones including opioids as the major constituents [1—5], and close interrelationships between opioids and melatonin have been proposed (for review see [2]).

In this paper, we aimed to review the experimental results, mainly from our work and related areas, on the adrenal medulla from the morphological point of view with special emphasis on its versatility and dynamic nature under normal day–night conditions and in relation to the effects of pinealectomy (PX), sham

Address for correspondence: T. Kachi, Department of Anatomy, Hirosaki University School of Medicine, Zaifucho 5, Hirosaki 036-8562, Japan.

PX (SPX) and melatonin. Differences and interrelationships between cell types and among various intracellular events are special issues. In our work, quantitative morphological studies were carried out using rats and golden hamsters.

Normal structure

Adrenaline cell and noradrenaline cell

In the adrenal medulla, there are at least two types of chromaffin cells, called adrenaline (A) cells which have PNMT activity and noradrenaline (NA) cells which have no PNMTactivity [6]. It is possible to differentiate these two types of cells by various methods [7]. The distribution pattern of A cells and NA cells is different among various species of mammals, as shown in Fig. 1 [8—12]. In particular, a marked difference exists between the rabbit type and the whale type. In the former, most of the catecholamines contained are E and in the latter, norepinephrine (NE), like in fetuses [13]. It is well known that the hyperglycemic and cardioacceleraive actions and the response to emotional stress are much higher in E than in NE. It is important to note that whales have large thyroid glands and small adrenal glands, and do not have well-developed pineal glands, olfactory bulbs, mammillary bodies, anterior thalamic nuclei, and hippocampi [14,15]. Because of the physiological adaptations in diving and temperature tolerance, whales show a decreased heartbeat for economy of the available oxygen during diving, and have a high metabolic rate. Another interesting issue is the distribution pattern in hamsters. In this species, NA cells exclusively distribute in the juxtacortical region [9,12]. Likewise, it has been reported that the adrenal glands of squamated (snakes) and the tuatara, *Sphenodon punctatus*, have peripherally located NA cells and central islets of A cells intermingled among adre-

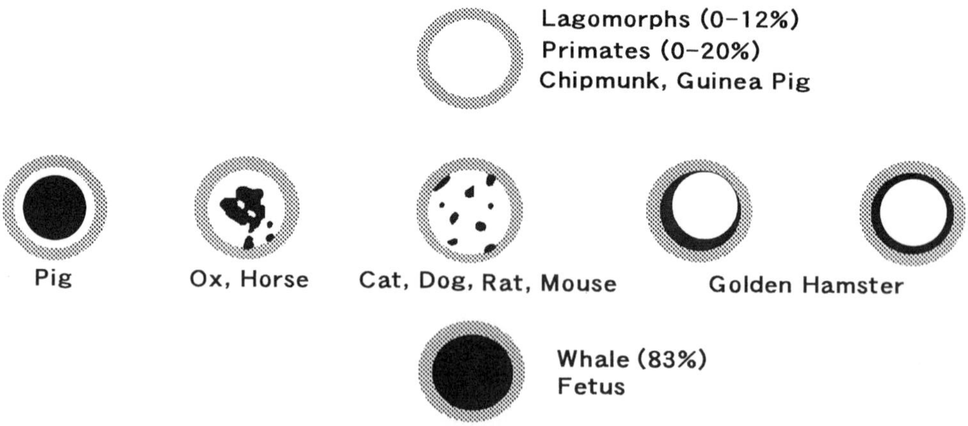

Fig. 1. Distribution pattern of A cells and NA cells in mammalian adrenal medullae. Clear, A cells; black, NA cells; stippled, inner cortex; parentheses, NE/total catecholamines. NA cells have been reported to exist in the adrenal medulla of man [10].

nocortical cells [11]. Thus, the biological diversity can be seen in the distribution pattern of A cells and NA cells and in the relative content of E and NE, and the differences in the mode of adaptation and the phylogenetic relation seem to be reflected in those, at least partially.

As shown in Table 1, there are many differences in morphological, biochemical and other characteristics between A cells and NA cells.

Regional difference, follicle and intracellular polarity

Although both A cells and NA cells distribute in the juxtacortical region in mice, a markedly higher rate of uptake of [^{3}H]dopa and other amines, probably caused

Table 1. Examples of differences between A cells and NA cells in rats and other mammals.

A > NA	A < NA
Chromaffin cell	
Cell – size, mitosis [16,17]	Number of subtypes (mouse) [19]
Gap junction – size, frequency [18]	Area covered with *supporting cells* containing nonspecific cholinesterase, S-100 protein[a] and GFAP [22,23]
Nucleus – size, heterochromatin, nuclear sap [20]	
Nucleus, nucleolus – 24-h changes [20,21]	Nucleolar margination [21]
Nucleolus – pars granulosa [20]	Secretory granules – reaction with GA, iodate oxidation at pH 5.0–6.0, electron density in GA·OsO_4-fixed material [5–7], largest granule [9], calcium concentration [26]
RER [54], Golgi area [9]	
Secretory granules – size uniformity, glutamate-rich soluble protein (ox, pig, sheep) [25], glycoprotein [10]	
Uptake – dopa, dopamine, NE, E, 5-HTP, 5-HT [5]	Histamine (see [4])
Acid phosphatase [8], GABA, GAD [27]	Interleukin-1 [29], spot 35 protein[a] [30], helodermine-like protein [31], neurotensin (cat), somatostatin (man-pheochromocytoma) [3]
Met- & Leu-enkephalins [4,28], CGRP, SP, VIP, NPY [4]	
Glycogen (mouse) [32]	
Nerve ending	
Nerve ending volume [33,34]	Synaptic profile density/cell (mouse, hamster)
Mitochondria – total volume, volume density [33,34]	Small clear vesicle density [32,35], large granular vesicle density [33,34]
Enkephalin [28]	Neurotensin (hamster) (see [4])
Others	
Secretion by hypoglycemia, acute hypoxia and passive emotional stress [36]	Secretion by cold, reserpine and aggressive reactions [36]
Postnicotine K^+-induced secretion (bovine cell) [37]	Nicotine-induced secretion (bovine cells) [37]
6-Hydroxydopamine effects (effects of supraspinal catecholaminergic pathway) [38]	5,7-Dehydroxytryptamine effects (effects of supraspinal serotonergic pathway) [38]

[a]Calcium-binding protein.

by ACTH, has been found in both types of cells in the juxtacortical zone compared with the central portion (see [5]). It should be mentioned here that the PX effect on the nucleolar size was prominent in A cells of the juxtacortical region in rats [21].

The occurrence of follicles in the adrenal medulla (see [39]) has been reported in hedgehogs, hamsters and oxes. In the hamster medulla, mature follicles are formed at about 20 days of age in the central A-cell region [39]. Both A cells and NA cells in hamsters show a greater degree of polarity than similar cells in rats [9]. It is interesting that the PX effect on the frequency of exocytoses was more marked in the perivascular region of plasma membrane than in the other membrane regions of the golden hamster A cell [40]. The existence of intercellular canalicules has been reported in the adrenal medulla of several animals and the follicular lumen was considered to be an enlarged intercellular space [39].

Nerve ending, invagination complex and supporting cell

The chromaffin cells of the adrenal medulla are innervated by preganglionic nerve fibers derived from the intermediolateral cell column in the thoracic spinal cord [5,41]. In these endings of the splanchnic nerves, at least two types of vesicles, a large number of small clear vesicles and a small number of large granular vesicles which exist in the deeper portion, are seen, in addition to mitochondria, synaptic membrane thickenings, coated pits and vesicles, filaments, and the occasional occurrence of multivesicular body [33—35]. The main neurotransmitter at the nerve terminal is acetylcholine. Peptides and NO have been known to act also as neurotransmitters or neuromodulators [3,5,42].

At least two shapes, small spherical and large elongated, sometimes ramified, can be differentiated in these nerve endings [35]. The former is a main type in the NA-cell region and appears as the type "invaginated in chromaffin cells" at a much lower rate in the A-cell region in hamsters. The latter is a common shape in rats and in the A-cell region in hamsters. It is likely that these nerve endings are changeable in their shape, size and probably position [43,44].

Cup-like to tube-like invaginations of the plasma membrane were frequently seen near the synaptic complex in the nerve endings on adrenomedullary chromaffin cells, including both A cells and NA cells, in rats and golden hamsters. Coated pits were frequently seen opening near the bottom of each invagination. Each invagination was accompanied by a corresponding club- or finger-shaped cytoplasmic protrusion of the chromaffin cell, or occasionally of the adjacent supporting cell. Generally, such a protrusion consisted merely of cytoplasmic matrix containing actin-like components. The combined set of structures, including the invagination, frequently accompanying coated pits in the nerve ending and its corresponding protrusion of the chromaffin cell or the supporting cell, was designated as the "invagination complex" (IC) [45]. It was considered that the IC is a characteristic and dynamic feature of the nerve terminals and that it may possibly have a role in a phenomenon such as recycling of synaptic vesicles.

In addition, a possibility remains that the IC is also related to the uptake (or release) of Ca^{2+}, peptides, and/or proteins into (or from) the nerve endings, especially into (or from) the deep portion. Significant time-of-day and neuroendocrine effects were demonstrable in these structures.

It is likely that the A-cell and NA-cell regions are different not only in the presence of E synthesizing activity but also in the structural framework. That is, in the parenchyma of adrenal medulla there is another type of cells, called supporting cells, in addition to A cells, NA cells and SGC (small granule chromaffin) cells [5—7]. Surface areas of chromaffin cells, and also of nerve axons and endings, surrounded by supporting cells, are larger in NA-cell regions than in A-cell regions [22].

In summary, characteristics of A cells include:

1) well-equipped for synthesis, degradation, and/or processing of E, peptides such as enkephalin, and related enzymes;
2) relatively active glucose-related pathways;
3) forming an amplifier system, that is, homogeneous cells connected with gap junctions without interruption by supporting cells;
4) stronger influence by humoral factors such as glucocorticoids [46,47]; and
5) autocrine or paracrine inhibitory control by opiate, ANF, and $GABA_B$ receptors agonists [48—50].

Nerve terminals and preterminal axons in A-cell regions show a relatively higher rate of denuded portions and the terminals tend to show irregular shapes. Characteristics of NA cells include:

1) heavier innervation;
2) division into small compartments by supporting cells;
3) high Ca^{2+}-storing ability; and
4) unique contents of secretory granules.

Thus the A-cell region appears to be closer to usual endocrine glands. Pineal influence was more apparent in the A-cell system than in the NA-cell system, and this will be mentioned later.

Effects of SPX, PX and melatonin

24-h changes

The quantity of morphological components in adrenomedullary chromaffin cells and preganglionic nerve endings on them showed 24-h changes in normal animals. Sizes of cell, cytoplasm, nucleus and nucleolus, the frequency of exocytoses and the number of synaptic vesicles are such examples [16,20,21,33,34,40]. SPX often suppressed these changes. On the other hand, these 24-h changes often increased, but sometimes decreased, in amplitude and showed modifications in patterns and phase shifts in PX animals [51]. The opposite effects of SPX vs. PX were most clearly shown in the nucleolar size of A cells and in the synaptic vesicle density in nerve endings of the rat adrenal medulla [21,33]. The possibility is suggested that 24-h variations in various components of A cells and in the melatonin secretion change in amplitude to the opposite direction to each other. In relation to this, the plasma melatonin level has been reported to increase after SPX in sea turtles (see [51]), and we also found unusually high plasma levels of melatonin in not all but several (about 30%) rats in the daily dark phase, suggesting a high amplitude pulsatile secretion of melatonin [43,52]. Another interesting point observed in PX animals is the discrepancy between A cells and NA cells in the responses (i.e., increase in A cells and decrease in NA cells) of nucleolar size and mitotic frequency, notably in the mid-dark phase [17,21].

Concerning the pineal influence on circadian rhythms, at least two mechanisms have been postulated: one is the neural mechanism via the suprachiasmatic nuclei having the melatonin receptor [53] and another is the calmodulin-melatonin binding mechanism existing in many types of cells [54,55]. The latter mechanism is especially interesting because marked differences in calcium-related mechanisms exist between A cells and NA cells (Table 1).

Effects of SPX and PX on the daily mean level or amplitude, or the mean level at specific time points of the day

Experimental results mainly from our work are briefly summarized in Tables 2 and 3. Morphological components in the adrenal medulla were divided into two types in the quantitative response to PX as compared with SPX. In the quantity such as the A-cell size and the frequency of exocytoses into the perivascular space [16,40], no difference was seen between normal and SPX animals. However, in the quantity of many other components such as those in nerve endings apparent changes were seen in SPX animals compared to normal animals, and changes due to PX were usually in the opposite direction to those due to SPX. The mechanism of such pineal-dependent SPX effects is currently unknown, but local, neural and/or endocrine responses to the intracranial-pressure changes, pain, and/or axotomy induced by opening the skull and cutting meninges (see [51] and Hökfelt in this forum) may be involved.

Table 2. Nerve ending plasticity — changes following pinealectomy. (Preganglionic nerve endings on A cells).

Rat (8 time points)

1. Small clear vesicle (density): SX < PX (dark phase) [33]
2. Shape — invagination complex (frequency): SX > PX [45]
3. 24-h rhythm — small clear vesicle (density) [33]
   SX: suppressed; PX: amplified, pattern-modified, phase-shifted

Golden hamster (L11h, D1h)

1. Small clear vesicle (density): SX < PX [56]
2. Size (nerve ending invaginated in A cell): SX < PX [44]
3. Position (frequency) — basal: SX > PX [43,44]
   invaginated in A cell + intercellular: SX < PX

Table 3. Two different responses in A cells following pinealectomy [2,24].

SPX < PX	SPX > PX
A: Tyrosine uptake, DBH activity (see [57,58]) Golgi smooth clear vesicle [61] Exocytosis[b][40] Mitochondria – number [56] Cell size [16] Mitosis [17] Nucleolus – pars granulosa [20], 24-h changes in size [21] (Free ribosome production)	B: Met-enkephalin level[a] [59,60] Golgi-coated vesicle [61] RER size [24] Glycogen level [43]

A: These chemical and morphological features showed no change or decreased[a] (B: increased) responses following SPX. [a]In SPX hamsters, melatonin reversed the effect of continuous lighting [60]. [b]Frequency of cell profiles showing exocytoses in perivascular-space-facing plasma membranes, and number of those exocytoses per cell.

On the other hand, it is surmised that the nerve endings in PX animals are in a functionally activated state or in a state adapted to such while those in SPX animals are in a suppressed state. Therefore, the increased frequency of IC is suggested to be an indicator of a chronically inhibited state of the nerve endings. In addition, it has been reported that the glucocorticoid secretion may be also suppressed in SPX rats (see [51]).

Intracellular compact cooperation hypothesis

As shown in Table 3, in PX animals, notably two types of changes in opposite directions were observed in fine structures and chemical events of A cells. That is, it is considered that the synthesis and secretion of E and their relating structures increase, and the degradation of catecholamine-synthesizing enzymes, the synthesis of methionine-enkephalin (met-ENK), and their relating structures decrease in PX animals compared to SPX animals. In addition, as shown in Table 4, an interesting relationship exists between amino acids involved in the E synthesis [62] and the amino acid sequence of met-ENK. Therefore, when the E synthesis (or met-ENK synthesis and processing) system is activated, the met-ENK synthesis and processing (or E synthesis) system is suppressed in A cells of PX (or SPX) animals. Thus the materials which became unneeded in one system are utilized in the other system in the same cell.

Table 4. Summary of "intracellular compact cooperation hypothesis"

- Amino acid sequence of methionine-enkephalin

Tyrosine - glycine - glycine - **phenylalanine** - **methionine**

- Epinephrine synthetic pathway

Phenylalanine $\rightarrow$ **Tyrosine** $\rightarrow\rightarrow\rightarrow$ NE $\xrightarrow[\text{S-adenosyl }\mathbf{methionine}]{\text{PNMT}}$ E

- Relative contents of secretory vesicles – SPX: E↓, met-ENK↑
PX: E↑, met-ENK↓

A similar shared mechanism exists between the membranes of RER and Golgi-coated vesicles, and the other membranes such as those of mitochondria, Golgi smooth vesicles, secretory vesicles, the nuclear membrane and the plasma membrane. Thus, more phospholipid molecules are utilized for the formation of former membranes in SPX animals and for the formation of latter membranes in PX animals.

The third example is the utilization of glucose. Experimental results indicated that more glucose molecules are utilized for the synthesis of large glycogen molecules, and probably also for GABA and lysosome enzymes, in A cells of SPX animals (Tables 1 and 3). On the other hand, the cell size and numbers of exocytoses and mitochondria increase in PX animals, showing that the energy demand in A cells becomes higher. From these, therefore, it is surmised that more glucose molecules are utilized via glycolysis and aerobic energy production pathway in A cells of PX animals. In addition, since A cells in PX animals seem to produce more ribosomes and secretory vesicles which contain not only amines and peptides but also nucleotides and polysaccharides [1,10], more glucose molecules are presumed to be utilized via the pentose phosphate shunt and for the synthesis of polysaccharides in these cells.

Such a dynamic nature of structures and chemical events in adrenomedullary A cells seems to be important for keeping intracellular mechanisms *compact* and *economical*, and additionally so for enabling A cells to play roles for *multi-purposes*; that is, A-cell hormones not only cope with emergency situations, causing the fight or flight reaction, but also may adjust the body's cells to the state or the level of activity fit for the protection against or adaptation to infection, tissue damage, etc. and to the recovery from those and the rest [63,64]. Furthermore, since the above-mentioned changes in A cells were observed in relation to the pineal functional states, it is likely that the functional activity level of A cells and/or the balance between E and met-ENK in their secretory granules can be influenced by the pineal hormone level. These experimental results give further support for the reality of the proposed close relationship between melatonin and the opioid system regarding analgesia [1,65], actions on immunocytes [66,67] and suppressive effects on sympathetic nerve activities [51].

Direct and indirect actions of melatonin on the adrenal medulla

Due to recent great progress on the melatonin receptor in the CNS [53] and our experimental data on preganglionic neurons innervating the adrenal medulla (Table 2), it is almost certain that the pineal action on the adrenal medulla is, at least in part, mediated via the neural mechanisms including supraspinal autonomic centers, in which possible examples of targets would be the hypothalamic SCN-PVN, the thalamic paraventricular nuclei and/or the area postrema-NTS [41,43,53]. However, direct and indirect actions of melatonin via the humoral route have also been suggested. Many experimental data indicate that the plasma levels of thyroid hormone, ACTH, glucocorticoid, and renin [68] (see [51]) are suppressed by melatonin or augmented by PX, and that melatonin has general antioxidative effects (see [1,2]). Melatonin also binds and activates an orphan of the nuclear receptor superfamily in various cells including PC 12 cells [69]. Guanylate cyclase-related mechanisms have been also postulated to be involved in the mechanisms of direct actions of melatonin (see [51]).

It has recently been shown that the activation of Ca^{2+}/ calmodulin-dependent protein kinase II induces physiologically important processes such as stimulation of catecholamine synthesis (substrate: tyrosine hydroxylase) and neurotransmitter release (synapsin I); and inhibition of actin filament gelation, microtubule polymerization (microtubule-associated protein) and glycogen synthesis (glycogen synthase) [54]. On the other hand, melatonin inhibits the calmodulin action and PX increases the plasma Ca^{2+} level [54,55] (see [51]). Therefore, it is surmised that many of our data in PX animals indicating the increased activity of neural transmission, the decreased occurrence of IC, the activation or its potentiality of E synthesis and secretion, and the breakdown of glycogen in A cells, can be interpreted to a certain extent as the diminution of a direct action of melatonin on the neuroendocrine complex of A cells.

Pineal influences were different between A cells and NA cells. This would be mostly due to the differences in the central mechanisms and their interrelationships [36,38,41], but a possibility remains that this is partially due to qualitative and quantitative differences between the A-cell system and the NA-cell system in the adrenal medulla. In this relation, it would be particularly important to note that the A-cell system is essentially an extensive amplifier system and more open to the humoral influence, and has a higher metabolic activity, proper inhibitory mechanisms, and a stronger relation to opioids while the NA-cell neuroendocrine complex is more tightly ensheathed by supporting cells and the NA cell-supporting cell complex has a higher ability for calcium storage (Table 1).

References

(We apologize that we could not cite many important original papers because of the page limitation.)

1. Quay WB, Kachi T. Amine-secreting endocrines. In: Timiras P, Quay WB, Vernadakis A (eds) Hormones and Aging. Boca Raton: CRC Press,1995;67—84.
2. Kachi T, Takahashi G, Suzuki T, Kimura N, Yanagisawa M, Kajihara S, Kudo H. Relationship between pineal gland and adrenal medulla. In: Tang PL, Pang SF, Reiter RJ (eds) A Universal Photoperiodic Signal with Diverse Actions. Basel: Karger. Front Horm Res 1996;21:51—59.
3. Hökfelt T, Johanson O, Ljungdahl Å, Lundberg JM, Schultzberg M. Peptidergic neurons. Nature 1980;284:515—521.
4. Kondo H. Immunohistochemical analysis of the localization of neuropeptids in the adrenal gland. Arch Histol Jpn 1985;48:453—481.
5. Coupland RE. The natural history of the chromaffin cells: Twenty-five years on the beginning. Arch Histol Cytol 1989;52(Suppl)9:331—341.
6. Coupland RE. Ultrastructural features of the mammalian adrenal medulla. In: Motta PM (ed) Ultrastructure of Endocrine Cells and Tissues. Boston: Martinus Nijhoff, 1984;168—179.

7. Böck P. The paraganglia. In: Oksche A, Vollrath L (eds) Handbuch der Mikroskopischen Anatomie des Menschen. Berlin: Springer, 1982;8–26,174–193.
8. Eränkö O. Fluorescing islets, adrenaline and noradrenaline in the adrenal medulla of some common laboratory animals. Ann Med Exp Biol Fenn 1955;33:278–290.
9. Wood JG. Identification of and observations on epinephrine and norepinephrine containing cells in the adrenal medulla. Am J Anat 1963;112:285–303.
10. Benchimol S, Cantin M. Ultrastructural cytochemistry of the human adrenal medulla. Histochemistry 1977;54:9–26.
11. Norris DO. Vertebrate Endocrinology. Philadelphia: Lea & Febiger, 1980.
12. Suzuki T, Kachi T. Similarities and differences in supporting and chromaffin cells in the mammalian adrenal medullae: an immunohistochemical study. Anat Rec 1996;244:358–365.
13. Lagercrantz H, Slotkin TA. The "stress" of being born. Sci Am 1986;254:100–107.
14. Slijper EJ. Whales: Hutchinson & Co., 2nd edn. 1979.
15. Vollrath L. The Pineal Organ. In: Oksche A, Vollrath L (eds) Handbuch der Mikroskopischen Anatomie des Menschen, VI/7. Berlin: Springer, 1981.
16. Kachi T, Banerji TK, Quay WB. Quantitative cytological analysis of functional changes in adrenomedullary chromaffin cells in normal, sham-operated and pinealectomized rats in relation to time-of-day. III. Nuclear density. J Pineal Res 1988;5:527–534.
17. Kachi T, Quay WB, Banerji TK, Imagawa T. Effects of pinealectomy on the mitotic activity of adrenomedullary chromaffin cells in relation to time of day. J Pineal Res 1990;8:21–34.
18. Grynszpan-Winograd O. Close relationships of mitochondria with intercellular junctions in the adrenalin cells of the mouse adrenal gland. Experientia 1982;38:270–271.
19. Gorgas K, Böck P. Morphology and histochemistry of the adrenal medulla. I. Various types of primary catecholamine-storing cells in the mouse adrenal medulla. Histochemistry 1976; 50:17–31.
20. Kachi T, Banergi TK, Quay WB. Quantitative cytological analysis of functional changes in adrenomedullary chromaffin cells in normal, sham-operated and pinealectomized rats in relation to time-of-day. II. Nuclear-cytoplasmic ratio, nuclear size and pars granulosa of nucleolus. J Pineal Res 1988;5:141–159.
21. Kachi T, Banerji TK, Quay WB. Quantitative cytological analysis of functional changes in adrenomedullary chromaffin cells in normal, sham-operated, and pinealectomized rats in relation to time of day. I. Nucleolar size. J Pineal Res 1984;1:31–49.
22. Kachi T, Suzuki T, Takahashi G, Quay WB. Differences between adreno-medullary adrenaline and noradrenaline cells: quantitative electron-microscopic evaluation of their differential cellular association with supporting cells. Cell Tissue Res 1993;271:257–261.
23. Suzuki T, Kachi T. Immunohistochemical studies on supporting cells in the adrenal medulla and pineal gland of adult rat, especially on S-100 protein, glial fibrillary acidic protein and vimentin. Acta Anat Nippon 1995;70:130–139.
24. Kachi T, Takahashi G, Banergi TK, Quay WB. Rough endoplasmic reticulum in the adrenaline and noradrenaline cells of the adrenal medulla: effects of intracranial surgery and pinealectomy. J Pineal Res 1992;12:89–95.
25. Hopwood D. Adrenal medullary basophilia in ox, pig and sheep. A histochemical, immunohistochemical and cell fractionation study. Histochemie 1967;11:268–279.
26. Daimon T. Electron microscopical microprobe analysis of chromaffin cells in rat adrenal medulla. Acta Anat Nippon 1993;68:662.
27. Häppölä O, Karhula T, Päivärinta H, Soinila S, Wu J-Y, Ahonen M. L-Glutamate decarboxylase immunoreactivity in the sympathoadrenal system. In: Erdö SL (ed) GABA Outside the CNS. Berlin: Springer, 1992;65–82.
28. Pelto-Huikko M, Salminen T, Hervonen A. Localization of enkephalins in cells and the nerves innervating adrenaline cells in rat adrenal medulla. Histochemistry 1985;82:377–383.
29. Schulzberg M, Anderson C, Unden A, Troye-Blomberg M, Svenson SB, Bartfai T. Interleukin-1 in adrenal chromaffin cells. Neuroscience 1989;30:805–810.

30. Kondo H, Kuramoto H, Iwanaga T, Fujita T. Cerebellar Purkinje cell-specific protein-like immunoreactivity in noradrenalin-chromaffin cells and ganglion cells but not in adrenaline-chromaffin cells in the rat adrenal medulla. Arch Histol Jpn 1985;48:421–426.
31. Bjartell A, Persson P, Absood A, Sundler F, Håkanson R. Helodermin-like peptides in noradrenaline cells of adrenal medulla. Reg Peptides 1989;26:27–34.
32. Costa JL. Histological characterization of "epinephrine" and "norepinephrine" cells in the adrenal medulla of the white-footed mouse, *Peromyscus leucopus*. Gen Comp Endocrinol 1973;20: 498–514.
33. Kachi T, Banerji TK, Quay WB. Daily rhythmic changes in synaptic vesicle contents of nerve endings on adrenomedullary adrenaline cells, and their modification by pinealectomy and sham operations. Neuroendocrinology 1979;28:201–211.
34. Kachi T, Banerji TK, Quay WB. Circadian and ultradian changes in synaptic vesicle numbers in nerve endings on adrenomedullary noradrenaline cells, and their modifications by pinealectomy and sham operations. Neuroendocrinology 1980;30:291–299.
35. Grynszpan-Winograd O. Adrenaline and noradrenaline cells in the adrenal medulla of the hamster: a morphological study on their innervation. J Neurocytol 1974;3:341–361.
36. Lewis GP. Physiological mechanisms controlling secretory activity of adrenal medulla. In: Blaschko H, Sayers G, Smith AD (eds) Handbook of Physiology 7. VI Adrenal Gland. Washington DC: American Physiological Society, 1975;309–319.
37. Marley PD, Livett BG. Differences between the mechanisms of adrenaline and noradrenaline secretion from isolated, bovine, adrenal chromaffin cells. Neurosci Lett 1987;77:81–86.
38. Lau C, Ross LL, Whitmore WL, Slotkin TA. Regulation of adrenal chromaffin cell development by the central monoaminergic system: differential control of norepinephrine and epinephrine levels and secretory responses. Neuroscience 1987;22:1067–1075.
39. Ito T. Postnatal histogenesis of the adrenal medulla of the hamster with special reference to the functional structure. Folia Anat Jap 1958;30:239–258.
40. Kachi T, Banerji TK, Quay WB. Quantitative ultrastructural analysis of differences in exocytosis number in adrenomedullary adrenaline cells of golden hamsters related to time of day, pinealectomy, and intracellular region. J Pineal Res 1985;2:253–269.
41. Loewy AD. Central autonomic pathway. In: Loewy AD, Spyer KM (eds) Central Regulation of Autonomic Functions. New York, Oxford: Oxford University Press, 1990;88–103.
42. Blottner D, Baumgarten H-G. Nitric oxide synthase (NOS)-containing sympathoadrenal cholinergic neurons of the rat IML-cell column: evidence from histochemistry, immunohistochemistry, and retrograde labelling. J Comput Neurol 1992;316:45–55.
43. Kachi T, Takahashi G, Suzuki T, Kimura N, Kajihara S, Kurushima M, Yamauchi T, Irie T, Kudou H, Saito Y, Hashimoto S, Honma K. Pineal effects on adrenal medulla, area postrema and brain water content in relation to intracranial surgery. Biol Signals 1997;6:255–263.
44. Yamauchi T. An Electron Microscopic Study on Nerve Endings on Adrenomedullary Adrenaline Cells in Golden Hamsters: Position, Shape and their Changes Due to Pinealectomy. Thesis, Hirosaki University, 1998.
45. Kachi T, Banerji TK, Quay WB. The invagination complex in nerve endings on adrenomedullary adrenaline cells: quantitative ultrastructural description, and analysis of changes with time-of-day and their modification by sham-surgery and pinealectomy. J Autonon Nerv Syst 1980;2:241–258.
46. Ciaranello RD. Regulation of phenylethanolamine N-methyltrasferase synthesis and degradation. I. Regulation by rat adrenal glucocorticoids. Molec Pharmacol 1978;14:478–489.
47. Unsicker K. The chromaffin cell. Paradigm in cell, developmental and growth factor biology. J Anat 1993;183:207–221.
48. Kumakura K, Karoum F, Guidotti A, Costa E. Modulation of nicotinic receptors by opiate receptor agonists in cultured adrenal chromaffin cells. Nature 1980;283:489–492.
49. Debinski W, Kuchel O, Buu NT. Atrial natriuretic factor is a new neuromodulatory peptide. Neuroscience 1990;36:15–20.

50. Oset-Gasque MJ, Castro E, Gonzalez MP. GABAergic mechanisms in bovine adrenal chromaffin cells: their role in the regulation of catecholamine secretion. In: Erdö SL (ed) GABA Outside the CNS. Berlin: Springer, 1992;167–181.
51. Kachi T. Pineal actions on the autonomic system. Pineal Res Rev 1987;5:217–263.
52. Kurushima M. Effects of Intracranial Surgery on Pineal Structures and Melatonin Secretion. Thesis, Hirosaki University, 1998.
53. Stankov B, Fraschini F, Reiter RJ. The melatonin receptor: distribution, biochemistry, and pharmacology. In: Yu HS, Reiter RJ (eds) Melatonin: Biosynthesis, Physiological Effects, and Clinical Applications. Boca Raton: CRC Press, 1993;155–186.
54. Antón-Tay F, Huerto-Delgadillo L, Ortega-Corona B, Benítez-King G. Melatonin antagonism to calmodulin may modulate multiple cellular functions. In: Touitou Y, Arendt J, Pévet P (eds) Melatonin and the Pineal Gland. Amsterdam: Elsevier, 1993;41–46.
55. Benítez-King G, Huerto-Delgadillo L, Antó-Tay F. Binding of [^{3}H]-melatonin to calmodulin. Life Sci 1993;53:201–207.
56. Kachi T, Banerji TK, Quay WB. Pineal adrenomedullary relations: hormonal mechanisms affecting tumor growth. In: Gupta D, Attanasio A, Reiter RJ (eds) The Pineal Gland and Cancer. London: Brain Research Promotion, 1988;333–344.
57. Csaba G, Nagy S. The regulatory role of the pineal gland on the thyroid gland, adrenal medulla and islets of Langerhans. Acta Biol Med Ger 1973;31:617–619.
58. Banerji TK, Quay WB. Effects of melatonin on adrenomedullary dopamine-β-hydroxylase activity in golden hamsters: evidence for pineal and dose dependencies. J Pineal Res 1986;3: 397–404.
59. Kachi T, Takahashi G, Suzuki T, Kimura N, Banerji TK, Quay WB. Effects of pineal and intracranial surgery on the adrenal medulla: quantitative morphological and immunohistochemical studies. In: Touitou Y, Arendt J, Pévet P (eds) Melatonin and the Pineal Gland: From Basic Science to Clinical Application. Amsterdam: Elsevier, 1993;277–280.
60. Kimura N, Kachi T. Effects of time of day, intracranial surgery and pineal hormone on methionine-enkephalin-like immunoreactivity in adrenal medullae of golden hamsters. Hirosaki Med J 1996;48:139–147.
61. Kajihara S, Takahashi G, Kachi T. Effects of pinealectomy and sham surgery on Golgi apparatus in adrenomedullary adrenaline cells. A quantitative electron microscopic study. Hirosaki Med J 1997;48:197–207.
62. Kirshner N. Biosynthesis of the catecholamines. In: Blaschko H, Sayers G, Smith AD (eds) Handbook of Physiology 7. VI Adrenal Grand. Washington DC: American Physiological Society, 1975;309–319.
63. North RA, Terry ME. Actions and distributions of opioid peptides in peripheral tissues. Br Med Bull 1983;39:71–75.
64. Bhargava HN. Opioid systems and immune functions. In: Plotnikoff NP, Murgo AJ, Faith RE, Wybran J (eds) Stress and Immunity. Boca Raton: CRC Press, 1991;329–342.
65. Lakin ML, Miller CH, Stoff ML, Winters WD. Involvement of the pineal gland and melatonin in murine analgesia. Life Sci 1981;29:2543–2551.
66. Fraschini F, Reiter RJ. Role of Melatonin and Pineal Peptides in Neuroimmunomodulation. New York: Plenum Press, 1991.
67. Maestroni GJM, Conti A, Covacci V. Melatonin-induced immuno-opioids: fundamentals and clinical perspectives. Adv Pineal Res 1994;7:73–81.
68. Motta M, Schiaffini O, Piva F, Martini L. Pineal principles and the control of adrenocorticotropin secretion. In: Wolstenholme GEW, Knight J (eds) The Pineal Gland. Edinburgh, London: Churchill Livingstone, 1971;279–301.
69. Becker-André M, Wiesenberg I, Schaeren-Wiemers N, André E, Missbach M, Saurat J-H, Carlberg C. Pineal gland hormone melatonin binds and activates an orphan of the nuclear receptor superfamily. J Biol Chem 1994;269:28531–28534.

Dynamic Cells: Cell Biology of the 21st Century
S. Yagihashi, T. Kachi and M. Wakui, editors.

Schwann cell precursors and signals that control Schwann cell generation

Rhona Mirsky and Kristjan R Jessen
Department of Anatomy and Developmental Biology, University College London, London, UK

Abstract. We have focused on the generation of cells of the Schwann cell lineage. The presence of mRNA for the myelin protein P_0 in migrating neural crest cells suggests that it is an early glial lineage marker. Previously, we identified an intermediate cell type in the Schwann cell lineage, the Schwann cell precursor, which differs in many respects from the immature Schwann cell. Schwann cell precursors are the major cell type in limb nerves at embryo day (E)14/15 (rat) and E12/13 (mouse). During a brief period they convert to immature Schwann cells and the β-neuregulins are the axonal signals required for survival and maturation to Schwann cells. β-neuregulins induce sustained high-level phosphorylation of the transcription factor CRE binding protein (CREB) in Schwann cells. Currently, we are identifying genes that regulate the transition from Schwann cell precursors to Schwann cells using differential mRNA fingerprinting methods.

Schwann cells themselves produce autocrine survival signals that regulate their survival in the absence of axons, a property important for nerve regeneration. These signals are absent in precursors and are unlikely to be neuregulins. We have found that Schwann cells express the signalling protein Desert hedgehog (Dhh) which shapes the formation of the perineurial and other connective tissue sheaths around peripheral nerves.

Keywords: desert hedgehog, myelin, neuregulins, P_0.

Introduction

Our general interest in peripheral nerve development has led us to focus on cells of the Schwann cell lineage. Most of the Schwann cells in adult nerves originate from the neural crest. The neural crest is probably not the only source of peripheral glia, however, since there is evidence that some Schwann cells in ventral roots arise from the ventral neural tube and that some satellite and/or Schwann cells of dorsal root ganglia originate from neuroepithelial cells in the spinal cord after migration of the neural crest is complete [1–3].

A major focus of our work is to determine how Schwann cells arise during embryonic nerve development, using rat and mouse nerves as model systems. The main points of interest include the initial choice of glial identity and generation of Schwann cell precursors by pluripotent neural crest cells and the two main developmental transitions in the lineage, i.e., the formation of immature Schwann cells from precursors and the induction of the two mature Schwann cell types, myelinating and nonmyelinating Schwann cells, from immature cells.

Address for correspondence: Rhona Mirsky, Department of Anatomy and Developmental Biology, University College London, Gower Street, London WC1E 6BT, UK.

This last step takes place essentially in the postnatal period, and is reversible, in contrast to the precursor to Schwann cell transition (Fig. 1).

The origin of the Schwann cell lineage

The neural crest is a transient population of cells that emerges from the dorsal part of the neural tube as it closes. The cells disperse, migrate away, proliferate and differentiate to form a variety of cell types, including neurons, glia, melanocytes, cartilage and smooth muscle. While some crest cells may already be committed to a particular fate, many crest cells can give rise to more than one differentiated cell type [4—8]. To understand how this happens, it is important to analyse how extrinsic signals control lineage choices. Two types of signals can be distinguished: instructive signals that affect or determine the lineage choice made by individual crest cells, and selective signals. The latter signals act to modify the eventual outcome of initial lineage choices by selectively promoting (or blocking) the survival, or proliferation of certain lineages. In crest development, it has proved easier to demonstrate the existence of selective signals than instructive ones. Examples of this type of signal include NT-3 which is likely to regulate proliferation of early cells in the sensory neuron lineage, endothelin 3 and steel factor which control the survival and proliferation of early cells in the melanocyte lineage, and β-neuregulins which govern survival of Schwann cell precursors [2,9,10] (see below).

Evidence of instructive signals in crest development so far comes only from in vitro studies. Thus, transforming growth factor β (TGF-β) promotes the generation of smooth muscle cells from crest cells, bone morphogenetic protein 2

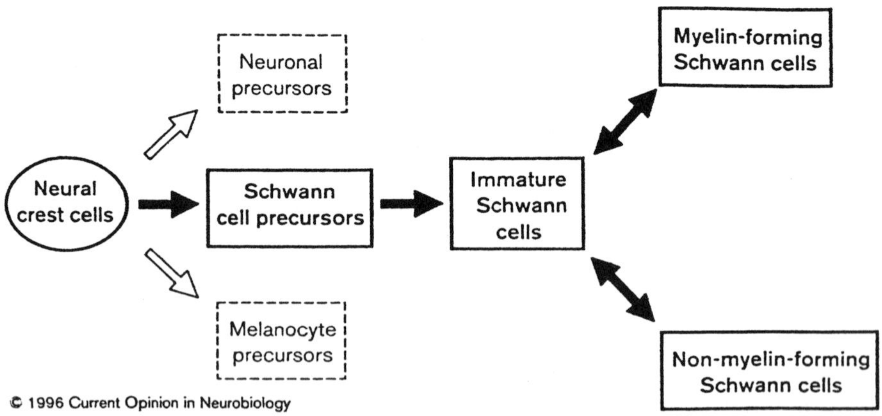

Fig. 1. A simplified outline of Schwann cell development and de-differentiation in the rat and mouse. Schwann cell precursors are a distinct cell type that generates cells with a recognizable Schwann cell phenotype (immature Schwann cell) before birth. This step appears to be irreversible. Further differentiation to form myelin-forming and nonmyelin-forming cells depends on axon-associated signals and is reversible (indicated by double-headed arrows). (Reproduced with permission from [17]).

(BMP2) stimulates entry to the autonomic neuronal lineage, and β-neuregulin blocks entry of crest cells to the neuronal lineage, thereby biasing crest differentiation towards the generation of glia [2]. This effect of β-neuregulin remains, however, to be reconciled with studies on β-neuregulin knockout mice which suggest that in the cephalic neural crest β-neuregulin is required for neurogenesis [11] (see below).

Early nerve formation

We know less about the formation of Schwann cell precursors from neural crest cells than we do about the generation of Schwann cells from precursors. Distinguishing between Schwann cell precursors and migrating neural crest cells is made more complicated by the fact that crest cells in the rat or mouse are almost certainly not a homogenous population, but are likely to be a mixture of multipotential cells and cells at various stages of entry to the different crest-derived lineages.

To make a nerve, prospective Schwann cells must find outgrowing axons and associate with them. The initial encounter between crest cells and axons probably takes place in the anterior parts of the somites that form along both sides of the neural tube. When neural crest cells migrate from the dorsal margin of the neural tube, many are directed through the anterior part of each somite, avoiding the posterior part. At approximately the same time, motor axons emerging from the ventrolateral area of the neural tube also project through the anterior part of each somite in preference to the posterior part. Therefore, both axons and crest cells are channelled to the same place at the same time. The molecular mechanisms that ensure this temporal and spatial matching of axon outgrowth and crest cell migration are not yet known. They appear to operate to a large extent independently of one another [3].

There is some evidence to suggest that crest cells enter the glial lineage quite early. Surprisingly, the evidence comes from observing the expression pattern of the gene for the myelin protein P_0. P_0 is the most abundant protein of Schwann cell myelin. Classically, it was considered to be a protein restricted to myelin, made as a part of a specific response to myelination-inducing axonal signals [12]. This picture must now be modified, since low, but clearly detectable, basal levels of P_0 are expressed not only in essentially all immature Schwann cells in the rat, irrespective of whether they will subsequently myelinate or not, but also in Schwann cell precursors [13]. The striking increase in P_0 synthesis that is selectively induced by axons in myelinating Schwann cells is therefore a strong upregulation of pre-existing basal levels rather than novel gene expression. Significantly, we found that P_0 is also present in a subpopulation of migrating neural crest cells in rodents, in agreement with findings in the chick [13,14]. There are no published reports of P_0 expression in cells other than Schwann cells and in view of this tight lineage-specificity of this gene in late development it is likely, although not proven, that P_0 expression in crest cells marks those cells that have just entered the Schwann cell lineage. It will be interesting to examine the relationship between P_0 expression in crest cells and expression of markers of entry to other crest lineages. If P_0 expression emerges as a specific marker of the glial lineage, it follows that many crest cells have selected a glial fate even at the early stages of crest migration.

The identification of Schwann cell precursors in rat and mouse nerves

In the development of the rat embryo, nerves first extend into the hind limb in significant numbers on embryo day (E)13/14. The first indication that the cells in these nerves differed significantly from Schwann cells in nerves of newborn animals came from the observation that cells dissociated from E14 nerves underwent programmed cell death within 20 h when they were plated at moderate/high density in vitro, conditions under which the Schwann cells survived well [15]. A more extensive comparison of these cells showed that E14 cells and newborn Schwann cells differed in many respects (Fig. 2). We concluded that the developing rat sciatic nerve harbored at least two distinct cell types during early development. The Schwann cell precursor is present in the nerve at E14, while by birth (E21—23) the nerve is populated by Schwann cells [9,15—17]. More recently, we have found that two days earlier, at E12, mouse nerves contain a Schwann cell precursor that is similar, albeit not identical, to the rat E14 precursor (Z. Dong, K.R. Jessen and R. Mirsky, unpublished).

In the rat sciatic nerve Schwann cells are generated from precursors at E15—17

To find out when Schwann cell precursors gave rise to Schwann cells, the phenotype of the cells in the sciatic nerve was determined on successive days between E14 and birth [15]. When we examined the ability of the cells to survive for 20 h in routine culture media in the absence of neurons we found that cells from E15

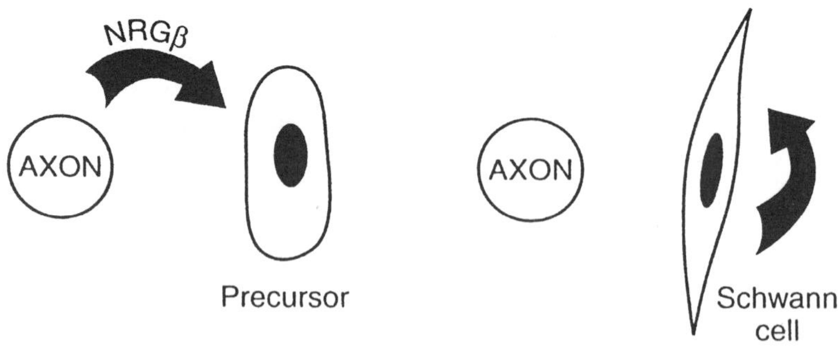

Fig. 2. Both axon-derived and autocrine signals regulate survival in the Schwann cell lineage. There is a shift from an exclusive reliance on axonal signals to the establishment of autocrine loops as precursors generate Schwann cells. The molecular nature of the autocrine signal is not known. NRGβ, n-β-neuregulin. (Reproduced with permission from [3].)

nerves died like E14 cells, while most cells from E17 nerves survived like newborn cells. Similarly, nearly all E15 cells expressed very low or undetectable cytoplasmic levels of the glial protein S100 while E17 cells, like newborn cells, contained readily detectable cytoplasmic S100. We also found that whereas rat Schwann cell precursors do not divide in the presence of basic fibroblast growth factor 2 (FGF-2), IGF-1 and forskolin, Schwann cells do. The mitogenic response to FGF-2 also appeared between E15 and E17. A shift in the expression of other transcription factors in in vitro morphology and cell-to-cell relationships in culture also occurs with a similar time course, as does significant expression of the zinc-finger protein Krox-20, a transcription factor of key importance in myelination [18].

The simplest view of these findings is that the nerve essentially contains precursors at E14 and E15 and Schwann cells from E17 onwards. The time between E15 and E17 is particularly important since during this period precursor cells progress to generate Schwann cells. This change in phenotype extends to many diverse and apparently unrelated cell features, indicating a major difference between the two developmental stages. Nevertheless, the transition occurs relatively abruptly, so that only during E16 does the sciatic nerve contain significant numbers of both cell types. This timing of Schwann cell generation is derived from studies on nerves in limb buds and growing limbs. Schwann cells in dorsal and ventral roots develop ahead of the schedule described here [18]. The generation of Schwann cells from precursors has now also been examined under well-defined conditions in culture (see below). These studies remove any doubt that the change in the phenotype of whole cell populations, as described above, represents conversions of individual precursor cells to Schwann cells.

While basal P_0 expression might provide the continuity that enables us to trace Schwann cell development as far back as the migrating crest (see above), differences are also evident between crest cells and precursors. One such difference is the relationship of these cells to their surrounding tissues, since, unlike the crest, one of the characteristic features of precursors is the intimate association they have with axons, forming the compact structure seen in early embryonic nerves. Furthermore, β-neuregulins do not appear to promote crest cell survival although they act as potent survival factors for Schwann cell precursors (see below) and precursors, unlike crest cells, express the protein GAP-43 [15].

β-neuregulin is the axonal signal that regulates precursor survival and differentiation

There is strong evidence that a single signal, β-neuregulin, regulates the transition from Schwann cell precursors to Schwann cells. Neuregulins (also known as NDF, heregulin, ARIA, GGF or SMDF) are generated from a single gene, exist both in membrane-bound and soluble isoforms and regulate growth and differentiation of several cell types in vitro [19]. As mentioned above, E14 rat Schwann cell precursors undergo programmed cell death when they are removed from embryonic nerves and placed in neuron-free cultures. Dorsal root ganglion (DRG) neurons secrete protein(s) that prevent this death, and isolated axonal membranes also possess activity that supports precursor survival [15] (N. Ratner, K.R. Jessen and R. Mirsky, unpublished). It is likely, therefore, that the survival of Schwann cell precursors in vivo is regulated by signals from the axons with which they associate. These signals also drive, or permit, the next step in the development of the lineage, since the precursors mature to generate Schwann cells in vitro in the presence of the neuron-derived proteins [9]. The neuronal signal can be mimicked by β forms of neuregulin, but not α forms. β-neuregulins support precursor survival in vitro for several days and, during this period, the cells alter their phenotype from that of precursors to that of Schwann cells. The time course of this Schwann cell generation in vitro is similar to that with which the glial population of developing nerves switches from the precursor phenotype to the Schwann cell phenotype in vivo. Most importantly, the action of the signal from DRG neurons that supports precursor survival and differentiation is blocked by a soluble hybrid protein containing the extracellular domain of the erbB4 receptor, a high affinity receptor for neuregulin that is not known to bind to any other growth factor [9]. Neuregulins are also present at the appropriate time and place in rat and mouse embryos: strikingly high expression of neuregulin mRNA is seen over the motor neurons of the ventral horn in the spinal cord and over DRG neurons, the two major sources of axons in embryonic peripheral nerves [20—23]. Taken together, these results strongly suggest that neuronally derived neuregulin is likely to act as a survival/maturation factor for Schwann cell precursors.

Experiments using knockout mice have strengthened our initial findings, and confirm a central role for neuronal neuregulin signalling in the development of Schwann cell precursors. Thus, neuregulin-deficient mice which die between E10.5 and 11.5 from heart defects, have remarkably few Schwann cell precursors associated with the dorsal and ventral roots at E10.5 [24]. Furthermore, mice deficient in the major neuregulin receptor erbB3 lack Schwann cell precursors at early stages of nerve development, and subsequently fail to develop Schwann cells in their peripheral nerve trunks [10], indicating that neuregulin signalling via the erbB3 receptor is vital for precursor development/survival. Analysis of transgenic mice that express type III neuregulins such as SMDF, but do not express neuregulins that contain an Ig domain in their extracellular domain (types I and II), indicates that expression of type III neuregulin in DRG and motor neurons is sufficient to support the development of Schwann cell precursors and Schwann cells [11].

Although Schwann cells have long been suspected of providing trophic support for peripheral neurons, substantive evidence for this has been scarce. The erbB3 knockout mice mentioned above have provided an important opportunity to test this idea, since they lack Schwann cell precursors and, consequently, Schwann cells, while satellite cells in sensory ganglia are apparently present. Examination of cell death and neuron numbers in these animals reveal that although DRG neurons are initially (E12) present in normal numbers, they show an abnormally high incidence of death already at E13 and by E18 about 80% of the neurons have died [10]. This indicates a relationship of mutual dependence between developing neurons and Schwann cells, in which the survival of Schwann cell precursors depends on neuron-derived β-neuregulin, while the survival of neurones depends on as yet unidentified precursor-derived survival signals (Fig. 2). It will be interesting to see whether this idea stands up to further tests and, if so, what the glial-derived survival signals are.

The ability to study Schwann cell generation from precursors in vitro, in the presence of neuregulin, should make it easier to analyse how this key step in Schwann cell development is regulated. It is already clear that precursors generate Schwann cells in a process that is largely independent of proliferation: precursor differentiation proceeds equally in cultures in which cell division is fast or slow, and precursors are capable of converting to Schwann cells without dividing at all [9]. Distinct signals are also being identified that, respectively, accelerate and delay Schwann cell generation. An interesting question that is still unresolved is whether the action of neuregulin on Schwann cell precursors is simply to block death or whether this signal also promotes differentiation. Thus it is possible that by ensuring survival, neuregulin permits the operation of an intrinsic differentiation programme in the precursor cells that dictates their conversion to Schwann cells. Alternatively, neuregulin might be required to drive this differentiation programme, in addition to its survival promoting function.

Neuregulin signalling involves phosphorylation of the cyclic AMP response element binding protein (CREB)

It is known that β-neuregulins bind to both erbB3 and erbB4 receptors with high affinity, and that these receptors, in combination with erbB2 initiate a cytoplasmic signalling cascade which results in DNA synthesis in both Schwann cell precursors and Schwann cells. Schwann cells possess receptors for erbB3, erbB4 and erbB2, and β-neuregulins phosphorylate these receptors [19]. The lack of Schwann cell precursors and later Schwann cells, observed in erbB3 knockout mice, suggests that the erbB3 receptor is vital for the survival of cells in this lineage. Until recently, however, little has been known about the molecular events downstream of receptor activation in Schwann cells, although it has been reported that β-neuregulin binding results in activation of ras and the MAP kinase pathway and transient elevation of intracellular cAMP levels about 12 h later [25]. We have found that β-neuregulin causes sustained phosphorylation of the transcription factor CREB on serine 133. CREB is expressed by both Schwann cell precursors and Schwann cells, with little change in level during development and differentiation [26]. The level of phosphorylation induced is higher and the response is more prolonged than that seen with forskolin, a classical activator of CREB via the protein kinase A pathway. Thus substantial phosphorylation above basal levels is seen within 5 min, is maximal by 30 min, and is still elevated at 1 h, declining significantly by 4 h. In contrast forskolin-induced phosphorylation is maximal at 10 min and has declined to basal levels within 1 h. The β-neuregulin-induced phosphorylation probably involves the MAP kinase pathway since incubation of Schwann cells with the MAP kinase inhibitor PD98059 significantly reduces the phosphorylation on serine 133. Incubation in the absence of Ca^{2+}, which abrogates CREB phosphorylation induced via the CaM kinase pathway, has no effect on β-neuregulin-induced phosphorylation [27].

Transcription factors involved in Schwann cell development

In contrast to our understanding of the Schwann cell lineage at the cellular level, relatively few genes that are definitively crucial to Schwann cell development have been identified. Studies using knockout mice indicate that both Krox-20 and Oct-6 (also called SCIP and Tst-1) are involved in myelination [18,28,29]. Knockout of Krox-20 results in mice with no peripheral myelin. The Schwann cells assume a 1:1 relationship with the larger axons and complete one-and-a-half spiral turns of membrane round the axon, indicating that they are arrested at the promyelin stage. Knockout of Oct-6 also results in either delay in or arrest of myelination with a similar phenotype in early postnatal stages to that seen in Krox-20 knockout mice. On the basis of indirect evidence it has also been suggested that Pax-3 may be involved in Schwann cell development [30]. Using differential mRNA fingerprinting methods we have attempted to identify genes that regulate the transition from Schwann cell precursors to Schwann cells. We have assumed that genes crucial to this transition may be characterized by rapid up- or downregulation at the mRNA level at critical transitional time points, and have developed a differential PCR strategy based on the modular structure of zinc-finger transcription factors, to design degenerate primers against consensus sequences. Total RNA derived from the developing mouse sciatic nerve on E12 and the first postnatal day was compared by RT-PCR. Using this approach we were able to identify a number of previously unknown zinc-finger transcription factors in the PNS. One of these zinc-finger genes has been cloned. The developmental expression of this factor in sciatic nerve and brachial plexus from E12 to adult life was compared with mRNA of other transcription factors expressed by cells of the Schwann cell lineage, namely Oct-6, Oct-1, Krox-20 and Pax-3 using RT-PCR [31] (G. Zoidl, K.R. Jessen, R. Mirsky, unpublished). The results showed that both the newly identified and known transcription factors examined show strong developmental regulation in embryonic nerves and that their expression patterns clearly differ from each other. They are thus likely to play distinct regulatory roles in the development of cells in the Schwann cell lineage.

How is Schwann cell survival regulated?

We and others have provided strong evidence that the survival of precursors depends acutely on axonal survival signals mediated by β-neuregulin (above). This arrangement is likely to have biological advantages since it would help to match glial numbers to axon numbers. It might also have a morphogenetic role by ensuring that precursors that strayed from axons to inappropriate locations away from the nerve would die. It would, however, be calamitous if the same arrangement applied to postnatal Schwann cells, since nerve damage and consequent loss of axonal contact would then threaten large-scale Schwann cell death, which in turn would remove the possibility of axonal regeneration after nerve damage. It might therefore be expected that postnatal Schwann cells possess an alternative, axon-independent, way of surviving. We have found that such a mechanism does indeed exist in the form of an autocrine survival loop. Importantly, such survival loops do not exist in Schwann cell precursors. In our initial comparison of survival of precursors and Schwann cells [15], the cells were grown at a density sufficient for Schwann cells to achieve a concentration of secreted survival factors that ensured survival. Precursors are unable to do this, irrespective of cell density, and, therefore, die in the absence of neuronal signals [15]. However, when neonatal Schwann cells are plated at extremely low cell density they no longer survive. In agreement with this, transection of adult nerves does not appear to lead to Schwann cell death, while degeneration of embryonic axons leads to death of precursors, at least in the chick [32–35]. Schwann cells in newborn nerves may be in an intermediate stage, having developed substantial axon-independent autocrine survival support, while partly relying on axonal survival signals [33,34]. We are at present investigating the identity of the autocrine Schwann cell growth factors. Although it has been reported that β-neuregulin can act as a survival signal for neonatal Schwann cells both in vivo and in vitro, and that Schwann cells can express β-neuregulin mRNA [36,37], we believe that the autocrine signal that we have identified is unlikely to be β-neuregulin. We already know that two characteristics of β-neuregulin do not apply to the autocrine Schwann cell activity: it does not cause Schwann cell DNA synthesis, and it will not keep Schwann cell precursors alive (C. Meier, R. Mirsky and K.R. Jessen, unpublished).

Desert hedgehog is involved in perineurium differentiation

Hedgehog (hh) molecules regulate important cell-to-cell interactions during embryonic development from *Drosophila* to mammals. In mouse, three Hedgehog genes have been cloned; Sonic hedgehog (Shh), Desert hedgehog (Dhh) and Indian hedgehog (Ihh) and hedgehog homologous have been isolated from other vertebrate species [38]. The most intensively studied of these genes is Shh, expressed in many locations in the embryo. It is involved in the specification of a variety of CNS neurons, including motor neurons, and influences the formation of several embryonic structures [38]. The main receptor for Dhh and Shh proteins is probably the transmembrane protein patched, acting with the protein smoothened.

Dhh has a more restricted distribution than Shh, being present in Sertoli cells, Schwann cell precursors and some endothelial cells during development. Apart from our studies the only known phenotype in Dhh knockout mice is male infertility, indicating a crucial role in spermatogenesis [39].

Our in situ and RT-PCR studies in collaboration with Prof Andrew McMahon, Harvard University, show that Schwann cells in developing mouse peripheral nerves express Dhh mRNA at E15. The mRNA is still strongly expressed at postnatal day 10 in myelin-forming Schwann cells but is downregulated in adult mouse nerves. Patched mRNA can be seen by in situ hybridisation in the mesenchymal tissue surrounding developing peripheral nerves at E15, a stage when the perineurial sheath is starting to organise around the nerve. When examined by electron microscopy, the connective tissue sheaths of the adult sciatic nerve of Dhh adult mice are malformed. In particular, the perineurium is strikingly reduced in thickness. Whereas the normal perineurial sheath consists of 6–8 layers of cells, in the mutant the cell layers vary from 2 to 4. Furthermore, the basal lamina is incomplete and the layers appear less compacted and taut than normal. The epineurium is reduced or absent, while numerous perineurial-like cells exist within the endoneurium, forming cell junctions and possessing incomplete basal lamina, a feature never seen in normal endoneurial fibroblasts. Thus these findings suggest an important role for Dhh, secreted by Schwann cells in signalling mesenchymal cells to organise the connective tissue sheaths around peripheral nerves (E. Parmantier, K.R. Jessen and R. Mirsky, unpublished).

References

1. Anderson DJ. Cell and molecular biology of neural crest cell lineage diversification. Curr Opin Neurobiol 1993;3:8–13.
2. Anderson DJ. Cellular and molecular biology of neural crest cell lineage determination. Trends Genet 1997;13:276–280.
3. Jessen KR, Mirsky R. Origin and early development of Schwann cells. Microsc Res Tech 1998; 41:393–402.
4. Frank E, Sanes JR. Lineage of neurons and glia in chick dorsal root ganglia: analysis in vivo with a recombinant retrovirus. Development 1991;111:895–908.
5. Fraser SE, Bronner-Fraser M. Migrating neural crest cells in the trunk of the avian embryo are multipotent. Development 1991;112:913–920.
6. Le Douarin N, Dulac C, Dupin E, Cameron-Curry P. Glial cell lineages in the neural crest. Glia 1991;4:175–184.
7. Stemple DL, Anderson DJ. Lineage diversification of the neural crest: in vitro investigations. Devel Biol 1993;159:12–23.
8. Weston JA. Sequential segregation and fate of developmentally restricted intermediate cell populations in the neural crest lineage. Curr Top Devel Biol 1991;25:133–153.
9. Dong Z, Brennan A, Liu N, Yarden Y, Lefkowitz G, Mirsky R, Jessen KR. NDF is a neuron-glia signal and regulates survival, proliferation, and maturation of rat Schwann cell precursors. Neuron 1995;15:585–596.
10. Riethmacher D, Sonnenberg-Riethmacher E, Brinkmann V, Yamaai T, Lewin GR, Birchmeier C. Severe neuropathies in mice with targeted mutations in the ErbB3 receptor. Nature 1997;389:725–730.

11. Meyer D, Yamaii T, Garratt A, Riethmacher-Sonnenberg E, Kane D, Theill LE, Birchmeier C. Isoform-specific expression and function of neuregulin. Development 1997;124:3575–3586.
12. Lemke G. Unwrapping the genes of myelin. Neuron 1988;1:535–543.
13. Lee M-J, Brennan A, Blanchard A, Zoidl G, Dong Z, Tabernero A, Zoidl C, Dent MAR, Jessen KR, Mirsky R. P_0 is constitutively expressed in the rat neural crest and embryonic nerves and is negatively and positively regulated by axons to generate non-myelin-forming and myelin-forming Schwann cells, respectively. Molec Cell Neurosci 1997;8:336–350.
14. Bhattacharyya A, Frank E, Ratner N, Brackenbury R. P_0 is an early marker of the Schwann cell lineage in chickens. Neuron 1991;7:831–844.
15. Jessen KR, Brennan A, Morgan L, Mirsky R, Kent AR, Hashimoto Y, Gavrilovic J. The Schwann cell precursor and its fate: a study of cell death and differentiation during gliogenesis in rat embryonic nerves. Neuron 1994;12:509–527.
16. Gavrilovic J, Brennan A, Mirsky R, Jessen KR. Fibroblast growth factors and insulin growth factors combine to promote survival of rat Schwann cell precursors without induction of DNA synthesis. Eur J Neurosci 1995;7:77–85.
17. Mirsky R, Jessen KR. Schwann cell development, differentiation and myelination. Curr Opin Neurobiol 1996;6:89–96.
18. Topilko P, Schneider-Manoury S, Levi G, Baron-Van Evercooren A, Chennoufi ABY, Seitanidou T, Babinet C, Charney P. Krox-20 controls myelination in the peripheral nervous system. Nature 1994;371:796–799.
19. Gassman M, Lemke G. Neuregulins and neuregulin receptors in neural development. Curr Opin Neurobiol 1997;7:87–92.
20. Ho W-H, Armanini MP, Nuijens A, Phillips HS, Osheroff PL. Sensory and motor neuron-derived factor. A novel heregulin variant highly expressed in sensory and motor neurons. J Biol Chem 1995;270:14523–14532.
21. Marchionni MA, Goodearl ADJ, Chen MS, Bermingham-McDonogh O, Kirk C, Hendricks M, Danehy F, Misumi D, Sudhalter J, Kobayashi K et al. Glial growth factors are alternatively spliced erbB2 ligands expressed in the nervous system. Nature 1993;362:312–318.
22. Meyer D, Birchmeier C. Distinct isoforms of neuregulin are expressed in mesenchymal and neuronal cells during mouse development. Proc Natl Acad Sci USA 1994;91:1064–1068.
23. Orr-Urtreger A, Trakhtenbrot L, Ben-Levy R, Wen D, Rechavi G, Lonai P, Yarden Y. Neural expression and chromosomal mapping of Neu differentiation factor to 8p12-p21. Proc Natl Acad Sci USA 1993;90:1867–1871.
24. Meyer D, Birchmeier C. Multiple essential functions of neuregulin in development. Nature 1995;378:386–390.
25. Kim HA, DeClue JE, Ratner N. cAMP-dependent protein kinase A is required for Schwann cell growth: interactions between the cAMP and neuregulin/tyrosine kinase pathways. J Neurosci Res 1997;49:236–247.
26. Stewart HJS. Expression of c-jun, Jun B, Jun D and cAMP response element binding protein by Schwann cells and their precursors in vivo and in vitro. Eur J Neurosci 1995;7:1366–1375.
27. Tabernero A, Stewart HJS, Jessen KR, Mirsky R. The neuron-glia signal β neuregulin induces sustained CREB phosphorylation on Ser-133 in cultured rat Schwann cells. Molec Cell Neurosci 1998;10:309–322.
28. Bermingham JR, Scherer SS, O'Connell S, Arroyo E, Kalla KA, Powell FL, Rosenfeld MG. Tst-1/Oct-6/SCIP regulates a unique step in peripheral myelination and is required for normal respiration. Genes Devel 1996;10:1751–1762.
29. Jaegle M, Mandemakers W, Broos L, Zwart R, Karis A, Visser P, Grosveld F, Meijer D. The POU factor Oct-6 and Schwann cell differentiation. Science 1996;273:507–510.
30. Kioussi C, Gross MK, Gruss P. Pax3: a paired domain gene as a regulator in PNS myelination. Neuron 1995;15:553–562.
31. Blanchard AD, Sinanan A, Parmantier E, Zwart R, Broos L, Meijer D, Meier C, Jessen KR, Mirsky R. Oct-6 (SCIP/Tst-1) is expressed in Schwann cell precursors, embryonic Schwann

cells, and postnatal myelinating Schwann cells: Comparison with Oct-1, Krox-20 and Pax-3. J Neurosci Res 1996;46:630–640.

32. Ciutat D, Calderó J, Oppenheim RW, Esquerda JE. Schwann cell apoptosis during normal development and after axonal degeneration induced by neurotoxins in the chick embryo. J Neurosci 1996;16:3979–3990.
33. Grinspan JB, Marchionni MA, Reeves M, Coulaloglou M, Scherer SS. Axonal interactions regulate Schwann cell apoptosis in developing peripheral nerve: neuregulin receptors and the role of neuregulins. J Neurosci 1996;16:6107–6118.
34. Syroid DE, Maycox PR, Burrola PG, Liu N, Wen D, Lee K-F, Lemke G, Kilpatrick TJ. Cell death in the Schwann cell lineage and its regulation by neuregulin. Proc Natl Acad Sci USA 1996; 93:9229–9234.
35. Trachtenberg JT, Thompson WJ. Schwann cell apoptosis at developing neuromuscular junctions is regulated by glial growth factor. Nature 1996;379:174–177.
36. Carroll SL, Miller ML, Frohnert PW, Kim SS, Corbett JA. Expression of neuregulins and their putative receptors, ErbB2 and ErbB3, is induced during Wallerian degeneration. J Neurosci 1997;17:1642–1659.
37. Raabe TD, Clive DR, Neuberger TJ, Wen D, DeVries GH. Cultured neonatal Schwann cells contain and secrete neuregulins. J Neurosci Res 1996;46:263–270.
38. Hammerschmidt M, Brook A, McMahon AP. The world according to Hedgehog. TIG 1997;13: 14–21.
39. Bitgood MJ, Shen L, McMahon AP. Sertoli cell signaling by Desert hedgehog regulates the male germline. Curr Biol 1996;6:298–304.

Dynamic Cells: Cell Biology of the 21st Century
S. Yagihashi, T. Kachi and M. Wakui, editors.

Supporting cells or glial cells in the adrenal medulla and pineal gland

Takao Suzuki, Gen Takahashi and Takashi Kachi
Department of Anatomy, Hirosaki University School of Medicine, Hirosaki, Japan

Abstract. Morphological and chemical nature of supporting cells in the adrenal medulla and glial cells in the pineal gland, and the effects of intracranial surgery on pineal glial cells were examined. Immunohistochemical or quantitative morphological results showed:
1. Extent of cellular association of chromaffin cells with supporting cells was remarkably higher in noradrenaline (NA)- than in adrenaline (A)-cell regions in many mammalian species, irrespective of proportions and distribution patterns of A cells and NA cells.
2. Distribution pattern and chemical nature of supporting and glial cells showed regional differences in both glands in rats. Several similarities were present in supporting and glial cells between in NA-cell regions and in the stalk and the proximal region of the pineal gland, and also between those in A-cell regions and in the distal region, respectively.
3. Relative lengths of glial cell membrane profiles adjoining the basal lamina of the perivascular spaces in the rat pineal gland were significantly decreased by intracranial surgery. Differences and similarities in cellular association of chief endocrine cells with supporting or glial cells may reflect those in biological and functional characteristics in the adrenal medulla and pineal gland, and their interaction and structural plasticity may be important in regulating activities.

Keywords: Cell-association pattern, chromaffin cell, immunohistochemistry, intracranial surgery, perivascular space, quantitative morphology.

Introduction

The adrenal medulla and pineal gland are the endocrine glands which originate from neuroectoderm, and contain at least two types of parenchymal cells, i.e., the chief amine-secreting type of cells (chromaffin cells and pinealocytes) and other type of cells (supporting cells in the adrenal medulla [1,2] and glial cells in the pineal gland [3,4]). These supporting and pineal glial cells have features similar to those of glial cells in the nervous system, and both glands have characteristics not only as endocrine glands, but also as nervous tissue [5–7].

The close cellular association of neurons and glia in the central nervous system and neuroendocrine system is a familiar notion, and neuron–glia interactions and their structural plasticity are considered to be important in regulating activities of both neurons and glia. In the adrenal medulla and pineal gland, therefore, it would be possible for similar interrelationships between chief endocrine cells and supporting cells or glial cells to be present. However, very little is known

---

Address for correspondence: T. Suzuki, Department of Anatomy, Hirosaki University School of Medicine, Zaifu-cho 5, Hirosaki, 036-8562 Japan.

regarding morphological interrelationships between these two types of parenchymal cells in both glands.

In this paper, we review our studies of the three following areas on the morphological relationship between chief endocrine cells and supporting cells or glial cells in the adrenal medulla and pineal gland.

Supporting cells in mammalian adrenal medullae

In many mammalian species, adrenomedullary chromaffin cells consist of adrenaline-chromaffin (A) cells, which have epinephrine-synthesizing enzyme (PNMT) activity, and noradrenaline-chromaffin (NA) cells, which have no PNMT activity. However, proportions and distribution patterns of these two types of chromaffin cells show species differences [2,8]. We attempted to elucidate whether common morphological characteristics and/or species differences are present in the interrelationships between supporting cells and chromaffin cells in both A- and NA-cell regions in the adrenal medullae of nine mammalian species (rat, golden hamster, chipmunk, rabbit, cat, dog, pig, ox, and horse) [9,10]. Cryostat sections of adrenal medullae fixed with Zamboni fluid for 24 h were immunostained by ABC methods using antisera to one of the glial marker proteins, S-100 protein [11,12], and PNMT.

S-100 protein-immunoreactive supporting cells were more numerously and densely distributed in NA-cell regions than in A-cell regions in seven species (rat, golden hamster, cat, dog, pig, ox, and horse), but not in the chipmunk and rabbit whose adrenomedullary chromaffin cells chiefly consist of A cells [2,10]. In NA-cell regions, these cells sent out fine cytoplasmic processes between NA cells and/or partially or largely enveloping the surface of NA cells, and forming a network by immunoreactive elements among NA cells. The frequency of S-100-labeled cells in the marginal portion of the parenchyma differed among species, but many S-100-labeled supporting cells were located in the inner portion of the parenchyma. The cell-association patterns of S-100-labeled supporting cells with NA cells were classified into three types [10]. In A-cell regions, relatively few S-100-labeled supporting cells were seen in most of the species including the chipmunk and rabbit, and these were located mostly in the marginal portion of the parenchyma. The frequency of S-100-labeled cells in A-cell regions were exceptionally higher in the horse than in the others.

The degree of cellular association of chromaffin cells with supporting cells was higher in the NA-cell region than in the A-cell region, as also shown by the quantitative electron microscopic study in rats [13]. These relationships were always seen in the adrenal medullae having both A cells and NA cells, irrespective of species differences in relative concentration of NA [14], and in volumetric proportions and/or distribution patterns of these two types of chromaffin cells [2,8]. Functional significance of these structural differences between the A-cell region and the NA-cell region have been discussed in a separate paper [2].

Regional differences of supporting cells and glial cells

The supporting cells and glial cells show an inhomogeneous distribution pattern in the adrenal medulla [9,10,13] or pineal gland [15–17]. It is interesting to examine whether there are regional differences not only in the morphology but also in the chemical nature of supporting and glial cells, and whether there are any relationships in such regional differences between these two glands. Cryostat sections of both glands of the rat were immunostained using antisera to the three glial marker proteins, S-100 protein, glial fibrillary acidic protein (GFAP), and vimentin.

In the adrenal medulla, supporting cells in NA-cell regions showed similar and intense immunoreactivities to GFAP and vimentin as well as to S-100 protein, forming a network by immunoreactive elements among NA cells. However, those in A-cell regions were less numerous and their immunoreactivity was positive to S-100 protein but negative to GFAP and, at least in part, negative to vimentin. In the pineal gland, glial cells in the stalk and the proximal region of the body portion showed immunoreactivities to the three antibodies, forming a network among pinealocytes. At the same time, those in the distal region of the body portion were less numerous and their immunoreactivity was positive to S-100 protein and vimentin, but negative to GFAP [18].

Both the adrenal medulla and pineal gland in the adult rat showed regional differences in the degree of cellular association of chief endocrine cells (chromaffin cells or pinealocytes) with, and in the expression of marker proteins in, supporting cells or glial cells. Comparing the morphological feature and chemical nature in supporting cells and glial cells, there are several similarities in these two cells between those in NA-cell regions of the adrenal medulla and in the stalk and the proximal region of the body portion of the pineal gland, and also between those in A-cell regions of the adrenal medulla and in the distal region of the body portion of the pineal gland, respectively. These two sets of corresponding regions in the two glands may have related functions or biological characteristics, respectively. At least, the nature of the former regions in both glands seems to be closer to that of nervous tissue than the latter regions which seem to be more typical endocrine tissues [1,4,13,19].

Effects of intracranial surgery on pineal glial cells

Concerning pineal effects on the adrenal medulla, our data indicate that the direction of sham pinealectomy (SPX) effects on morphological components of the adrenal medulla is usually opposite to that of pinealectomy (PX) effects [20–23]. Therefore, we sought to clarify electron microscopically, whether and how, morphological features of glial cells in the pineal gland are influenced by SPX.

Male Wistar rats (NO, n = 3; SPX, n = 3) at 7 weeks of age, 2 weeks after surgery, maintained under LD 14/10 were sacrificed at 4 h after the onset of darkness. Pineal glands were fixed with 2.5% GA and postfixed with 1% OsO_4, and processed following conventional EM methods. Montage photographs of 38 perivascular spaces of normal animals and 33 of SPX animals were used for detailed observation. Considering the report on changes in the perivascular regions after adrenalectomy [24], we measured the length and number of glial elements contacting the perivascular space in the distal region of the body portion of the pineal gland.

Glial cells and their cytoplasmic laminar processes showed markedly electron-dense appearances partly due to their richness in intermediate filaments and were observed near to, or in contact with, the perivascular space. In the normal rat, the pinealocytes were separated from the perivascular space by the barrier of glial cells and/or their cytoplasmic processes and the pinealocyte–perivascular contact area was narrow. The relative length of glial cell membrane profile to total length of pinealocyte and glial cell membrane profiles facing the basal lamina of the perivascular space was 58.5 ± 3.1% (mean ± standard error); in the SPX rat this value significantly decreased to 43.6 ± 3.0% (t test: $p < 0.002$). That is, after the SPX treatment glio-perivascular contact area diminished and the degree of exposure of pinealocytes to the perivascular space increased by almost 15%. The relative numbers of glial cells (and/or terminals of glial cell processes) and pinealocytes (and/or pinealocyte processes) abutting to the perivascular space were 47.1 ± 1.9% and 52.9 ± 1.9% in the normal rat, and 39.5 ± 2.0% and 60.5 ± 2.0% in the SPX rat, respectively. The difference in numerical ratio of glial cells between the normal and SPX rats was significant ($p < 0.01$).

SPX induces a decrease in the glio-perivascular contact area in the pineal gland of the rat. Similar changes have been reported in adrenolectomized rats [24]. Physiological and/or physical stimuli induce neuron–glia morphological changes in several neuroendocrine systems other than the pineal gland, and lead to an increased neuro-vascular contact area [25—29]. These and our results indicate the following possibility: that the degree of exposure of pinealocytes to the perivascular space increases under the state of increased functional activities of the pineal gland. Supporting evidence has been reported in sea turtles that the plasma melatonin levels increase in SPX animals [30]. Similar trends suggesting pulsatile secretion have been observed in rats [31].

With regard to the possible functional significance of glial plasticity, it is surmised that the enlarged exposure of pinealocytes to the perivascular space in SPX animals would facilitate pinealocytes not only to release hormonal substance(s) into, but also to receive influences of neural, hormonal and other local factors from, the perivascular space. Although it is postulated that the retraction of glial cell processes and enlargement of pinealocytes and/or their cytoplasmic processes in SPX animals, much remains to be done, also about details of morphological plasticity in the surroundings of pineal capillaries following the SPX treatment.

Acknowledgements

We wish to thank Mr T. Koizumi and Mr A. Hirata for their technical assistance. This work was supported in part by The Karoji Memorial Fund for Medical Research in Hirosaki University, Japan, granted to T. Kachi.

References

1. Coupland RE. Ultrastructural features of the mammalian adrenal medulla. In: Motta PM (ed) Ultrastructure of Endocrine Cells and Tissues. Boston: Martinus Nijhoff, 1984;168–179.
2. Kachi T, Takahashi G, Suzuki T, Kimura N, Kajihara S, Kurushima M, Yamauchi T, Irie T. Dynamic aspects of adrenomedullary structures in relation to pineal and intracranial surgery. In: Yagihashi S, Kachi T, Wakui M (eds) 1st Hirosaki International Forum of Medical Science: Dynamic cells: Cell Biology of the 21st Century. Amsterdam: Elsevier Science, 1998;47–58.
3. Vollrath L. The pineal organ. In: Vollrath L, Oksche A (eds) Handbuch der Mikroskopischen Anatomie des Menschen VI/7. Berlin: Springer, 1981;177–180.
4. Pévet P. Anatomy of the pineal gland of mammals. In: Relkin R (ed) The Pineal Gland. New York: Elsevier Science, 1983;1–62.
5. Relkin R. The Pineal Gland. Amsterdam: Elsevier, 1983.
6. Coupland RE. The natural history of the chromaffin cells: Twenty-five years on the beginning. Arch Histol Cytol 1989;52(Suppl):331–341.
7. Quay WB, Kachi T. Amine-secreting endocrines. In: Timiras P, Quay WB, Vernadakis A (eds) Hormones and Aging. Boca Raton: CRC press, 1995;67–84.
8. Eränkö O. Fluorescing islets, adrenaline and noradrenaline in the adrenal medulla of some common laboratory animals. Ann Med Exp Biol Fenn 1955;33:278–290.
9. Suzuki T, Kachi T. Differences between adrenaline cells and noradrenaline cells in cellular association with supporting cells in the pig adrenal medulla: an immunohistochemical study. Neurosci Lett 1994;176:217–220.
10. Suzuki T, Kachi T. Similarities and differences in supporting and chromaffin cells in the mammalian adrenal medullae: an immunohistochemical study. Anat Rec 1996;244:358–365.
11. Moore BW. A soluble protein characteristic of the nervous system. Biochem Biophys Res Commun 1965;19:739–744.
12. Cocchia D, Michetti F. S-100 antigen in satellite cells of the adrenal medulla and the superior cervical ganglion of the rat: an immunochemical and immunocytochemical study. Cell Tis Res 1981;215:103–112.
13. Kachi T, Suzuki T, Takahashi G, Quay WB. Differences between adrenomedullary adrenaline and noradrenaline cells: quantitative electron microscopic evaluation of their differential cellular association with supporting cells. Cell Tis Res 1993;271:257–261.
14. Holzbauer M, Sharman DF. The distribution of catecholamines in vertebrates. In: Blaschko H, Muscholl E (eds) Catecholamines. Berlin: Springer, 1972;110–185.
15. Møller M, Ingild A, Bock E. Immunohistochemical demonstration of S-100 protein and GFA protein in interstitial cells of rat pineal gland. Brain Res 1978;140:1–13.
16. López-Muñoz F, Calvo JL, Boya J, Carbonell AL. Coexpression of vimentin and glial fibrillary acidic protein in glial cells of the adult rat pineal gland. J Pineal Res 1992;12:145–148.
17. Borregón A, Boya J, Calvo JL, López-Muñoz F. Immunohistochemical study of the pineal glial cells in the postnatal development of the rat pineal gland. J Pineal Res 1993;14:78–83.
18. Suzuki T, Kachi T. Immunohistochemical studies on supporting cell in the adrenal medulla and pineal gland of adult rat, especially on S-100 protein, glial fibrillary acidic protein and vimentin. Acta Anat Nippon 1995;70:130–139.
19. Korf H-W, Sato T, Oksche A. Reflections on the nature and connectivities of pinealocytes. In:

Reiter RJ, Pang SF (eds) Advances in Pineal Research 3. London: John Libbey & Co Ltd., 1989; 11–16.

20. Kachi T, Banerji TK, Quay WB. Quantitative ultrastructural analysis of differences in exocytosis number in adrenomedullary adrenaline cells of golden hamsters related to time of day, pinealectomy, and intracellular region. J Pineal Res 1985;2:253–269.
21. Kachi T, Banerji TK, Quay WB. Quantitative cytological analysis of functional changes in adrenomedullary chromaffin cells in normal, sham-operated, and pinealectomized rats in relation to time of day: II. Nuclear-cytoplasmic ratio, nuclear size and pars granulosa of nucleolus. J Pineal Res 1988;5:141–159.
22. Kachi T, Takahashi G, Suzuki T, Kimura N, Yanagisawa M, Kajihara S, Kudo H. Relationship between pineal gland and adrenal medulla. Front Horm Res 1996;21:51–59.
23. Kachi T, Takahashi G, Suzuki T, Kimura N, Kajihara S, Kurushima M, Yamauchi T, Irie T, Kudou H, Saito Y, Hashimoto S, Honma K. Pineal effects on adrenal medulla, area postrema and brain water content in relation to intracranial surgery. Biol Signals 1997;6:255–263.
24. Deußen-Schmitter M, Garweg G, Schwabedal PE, Wartenberg H. Simultaneous changes of the perivascular contact area and HIOMT activity in the pineal organ after bilateral adrenalectomy in the rat. Anat Embryol 1976;149:297–305.
25. Wittkowski W, Brinkmann H. Changes of extent of neuro-vascular contacts and number of neuro-glial synaptoid contacts in the pituitary posterior lobe of dehydrated rats. Anat Embryol 1974;146:157–165.
26. Wittkowski W, Scheuer A. Functional changes of the neuronal and glial elements at the surface of the external layer of the median eminence. Z Anat Entwickl-Gesch 1974;143:255–262.
27. Tweedle CD, Hatton GI. Evidence for dynamic interactions between pituicytes and neurosecretory axons in the rat. Neuroscience 1980;5:661–667.
28. Luckman SM, Bicknell RJ. Morphological plasticity that occurs in the neurohypophysis following activation of the magnocellular neurosecretory system can be mimicked in vitro by β-adrenergic stimulation. Neuroscience 1990;39:701–709.
29. Theodosis DT, MacVicar B. Neurone–glia interactions in the hypothalamus and pituitary. Trends Neurosci 1996;19:363–367.
30. Owens DW, Gern WA. The pineal gland and melatonin in sea turtles. In: Lofts B, Holmes WN (eds) Current trends in comparative endocrinology. Hong Kong: Hong Kong Univ Press, 1985; 645–648.
31. Kurushima M, Kachi T, Hashimoto S, Honma K. Comparative studies on plasma melatonin concentrations and pineal lipid contents: effects of time of day and various intracranial surgeries. Acta Anat Nippon 1997;72:341.

Dynamic Cells: Cell Biology of the 21st Century
S. Yagihashi, T. Kachi and M. Wakui, editors.

Immunological kinship between retinal and pineal photoreceptor cells in Lamprey (*Lampetra japonica*)

Akira Tonosaki and Keiji Mori
Department of Anatomy, Yamagata University School of Medicine, Yamagata, Japan

Abstract. Pineal and parapineal bodies of adult river lamprey (*Lampetra japonica*) were studied by immunohistochemistry. The floor of the pineal end-vesicle was elongate in shape, 600 μm anteroposteriorly, 370 μm transversely, and consisted of a centroposterior elevated region with an ostium descending into the atrium. The peripheral region showed a shallow concavity to reach the side wall. Serial frozen sections were placed in contact with autologous monoclonal antibodies, S70, specific for the outer segment of the short photoreceptor (rod type) of the retina; L31, for that of the long cell (cone type); and H16, for the outer segment of both types of cells. S70 showed intense reaction in the broad central region of the floor of the pineal end vesicle. In contrast, L31's positive reaction was limited to the outer segments projecting in a narrow peripheral region, almost offset to that of S70. H16 stained the outer segments of the floor cells entirely. Of the atrium, however, cells were stained with either S70 or L31. Consequently, luminal epithelia of the end-vesicle were differentiated into the central rod-type- and peripheral cone-type-dominant regions. It is noteworthy that, in the retina, rod- and cone-type cells are distributed evenly by a ratio of 3:1 all over the regions. Cells of the floor of the parapineal vesicle showed positive reaction with H16, S70 as well as L31. Such regional division as visualized here seems to correspond with reported results from application of heterologous or biotechnical rod-opsin and iodopsin immunoreactivity. There is an evolutionary parallelism between retinal and pineal photoreceptor cells.

Morphology of photoreceptor cells, historical

Vertebrate photoreceptor cells have a unique polarized structure. As early as the 1870s, the histology of the human retina, consisting of rod- (short photoreceptor) and cone-(long cell)-type photoreceptor cells, was established and illustrated with amazing precision [1]. The origin and location of photoreceptor cells were described in relation to a coronary section of the optic cup connected with the diencephalon via the stalk in a human embryo [2]. Ependymal cells in contact with cerebrospinal fluid transform into such cells. Going back to the origin of the optic cup, the anlagen emerge out of the cephalic corner ("Augengrube") of the neural plate [2].

Since von Frisch [3] and Scharrer [4], several bioscientists have investigated ichthyic and amphibian species after resection of eyes and pineal complex, and have developed an important theory — that photoreceptors would not be limited to the retina, but also occur in the other hypothalamic regions, particularly, in contact with cerebrospinal fluid [5]. Only lately, García-Fernández and collabora-

Address for correspondence: Akira Tonosaki, Department of Anatomy, Yamagata University School of Medicine, Yamagata 990-9585, Japan.

tors [6,7] have demonstrated immunohistochemical evidence of expression of phototransduction-related proteins in several heterotopical sites of the central nervous system of lampreys (*Lampetra japonica*). They predicted that such cells would be identified electrophysiologically in future.

The evidence for the ciliary origin of rod and cone outer segments is preserved in their connecting cilium joining the outer with the inner segment [8]. The connecting cilium has a set of centrioles but lacks central filaments. With freeze-fracturing, plasma membrane of the ciliary shaft shows a special necklace pattern of intramembranous particles, presumably corresponding to active transport of membrane material to the highly consuming outer segment.

Rod, and cone, outer segment contains compiled stacks of membranous discs which actually formed by infolding plasma membrane in the most proximal part of the outer segment. It is plausibly accepted that the continuation of discs to plasma membrane is lost shortly after formation in the proximal part of rod outer segment; meanwhile it is retained to relatively distal parts in the case of the cone outer segment.

Lamprey, a key to the evolution of vertebrate photoreceptor cells

Lamprey is one of several extant agnatha, or jawless, species. The genus is believed to be a 500-million-year-old descendant from Ordovician placodermi or the oldest ancestors of vertebrates. Our subject was about 40—50 cm in length, about 200—300 g in body weight, and regarded as a very close relative to *L. fluviatilis* and *Petromyzon marinus*.

On the surface or apical area of the neuromast, epidermal cells have a central cilium and surrounding microvilli, which are slender compared with the supporting cells [9]. Similar composition out of a cilium and significantly slender microvilli is generally observed on the body surface of mammalian embryos in contact with amnion [10]. The cilium is known as a common origin of photoreceptors, rod and cone outer segments; meanwhile slender microvilli will develop "calycal processes".

Ocular retina of lamprey

The retina is characterized by the bilayer formation of long and short photoreceptor cells (Fig. 1). The microscopic view is enhanced by respective ellipsoid bodies which are stained distinctively dark with toluidine blue. The long cell's length is accounted for by its very long myoid process. Its outer segment is rather short in size and placed in contact with the pigment epithelium. The short cell has a longer outer segment and its ellipsoid bodies are lined up horizontally above the external limiting membrane. The long cell outer segment is as a rule inserted into the pigment epithelial cytoplasm; the short cell outer segment barely reaches the tip of the process.

Based on amino acid sequencing, Okano et al. [11] classified the rod-opsin iso-

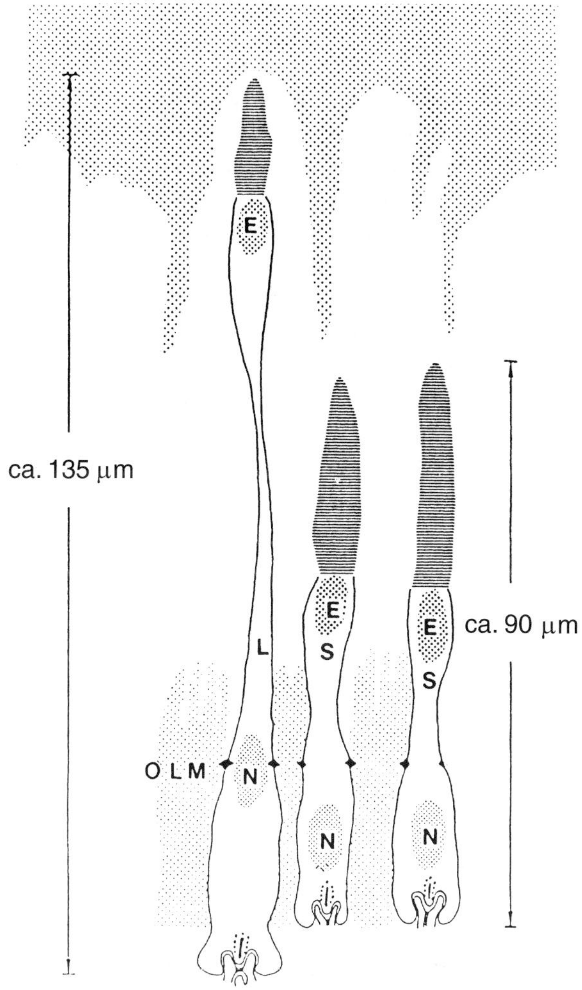

Fig. 1. Diagram showing the profiles of long and short photoreceptor cells of lamprey retina. Note that the outer segment of the long cell is smaller in size and inserted up into the pigment epithelial cytoplasm; the outer segment of the short cell, by contrast, is larger and takes place at a lower level of the "rod and cone layer". The long cell is also characterized by its long myoid and axonal processes exceeding its counterparts in the short cell. Redrawn after Ishikawa et al. [12].

lated from lamprey as the most primitive one. The long cell or cone-type cell, however, seems to contain other unresolved type(s) of pigment(s).

Immunohistochemical approach to lamprey photoreceptor cells

Ishikawa et al. [12] applied antibovine rhodopsin antibody to lamprey retina to obtain a clear-cut discrimination of the short against the long cell outer segment. According to them and a contemporary study by Negishi et al. [13], it is very likely that the short cell corresponds to the rod cell, and the long cell to the cone cell of higher vertebrates.

There is a criterion to distinguish the rod from the cone outer segment: this is that membranous discs are in connection with the outer plasma membrane. Counting such connection on TEM micrographs of dark- and light-adapted subjects, it turned out that the long and short cells have similarly 10—13% of discs in connection with the outer plasma membrane [14].

In the outer plexiform layer, on the other hand, the long cell axon terminal becomes more *ad vitreous* than the short cell terminal, quite similar to the case of cone pedicle in higher vertebrate species (Fig. 1). Characteristic ribbon synapses are also well developed in both terminals [15]. Various monoclonal antibodies against lamprey retinal homogenates have been raised by the authors and collaborators. Beginning with a group of antibodies, H16 and B11, which recognized outer segments of both long and short cells, we obtained antibodies including S70 and L31 which discriminate outer segments of each cell type [16,17] (Fig. 2).

Pineal and parapineal bodies of Lamprey

Since 1960s the morphology and fine structure of the pineal and parapineal bodies have been studied in order to classify photoreceptor cells type 1 and type 2. Type 1 is a columnar cell with an outer segment exactly recognized by anti-rod-opsin antibodies, presumably, therefore, specialized as a rod-like photoreceptor (Fig. 3). Type 2 is more or less a developed form of outer segment, frequently referred to as cone-related photoreceptor and halfway endocrine.

By immunohistochemistry with these antiphotoreceptor antibodies, we have been analyzing the nature and topography of pineal and parapineal luminal epithelia. Pineal body of lamprey is about 500 μm in size, situated approximately above and left to the parapineal body (Fig. 4).

We applied antibodies H16, which recognize both types of outer segments in

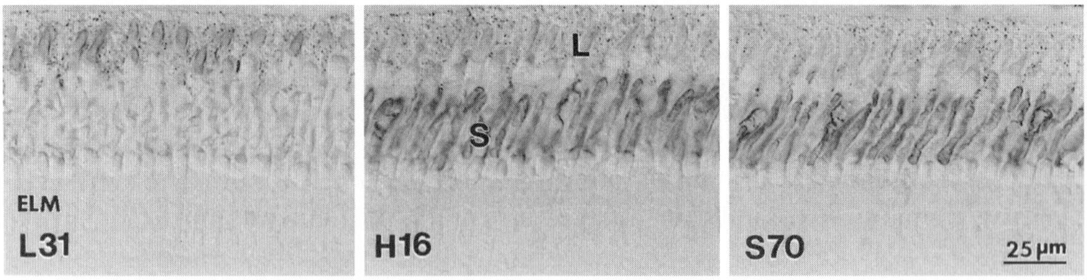

Fig. 2. Differential interference microscopy of serial trio out of radial frozen section of lamprey retina. Note the external limiting membrane at the bottom, the pigment epitheoium trimmed at the top. Diaminobenzidine reaction shows effects of antibodies: L31, specific for the outer segment of long cell; H16, specific for both outer segments, more intense for that of short cell; and S70, specific for the outer segment of short cell and, barely traceably, for its cytoplasm. ELM, external limiting membrane; L, long (cone type) cell; S, short (rod type) cell.

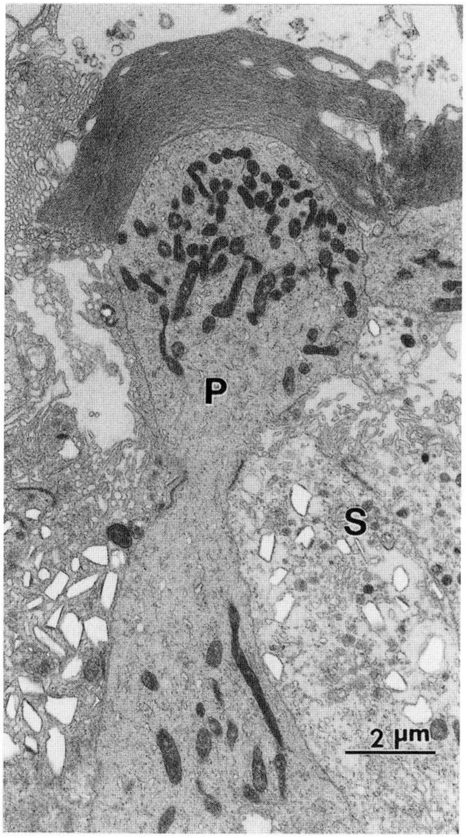

Fig. 3. TEM micrograph of a longitudinal section of the type 1 photoreceptor cell out of the pineal floor. Note the stack of membranous discs of the outer segment. In the pineal floor, pigment epithelial cells stand side by side with photoreceptor cells, not face-to-face across the subretinal space as in the case of the retina. Type 2 cell is not shown. P, photoreceptor cell with the outer segment; S, supporting cell (with melanin granules and lucid spaces after extraction of crystaloid substance).

the retina, L31 only positive for the long cell, and S70 positive for the short cell outer segment, to alternate serial sections. Intense "diaminobezidine reaction" was obtained with H16 and by S70 in the cenral region of floor epithelum, while with L31 a positive reaction was limited to the periphery of the floor of the pineal body (Fig. 4). In more posterior sections, H16 and S70 positive reaction was followed down to halfway of the pineal stalk and presumably up to epithalamic regions.

Conclusion

The long or cone type of the retina represents an ancient photoreceptor; the short or rod type is specialized according to the molecular pedigree of visual pigments

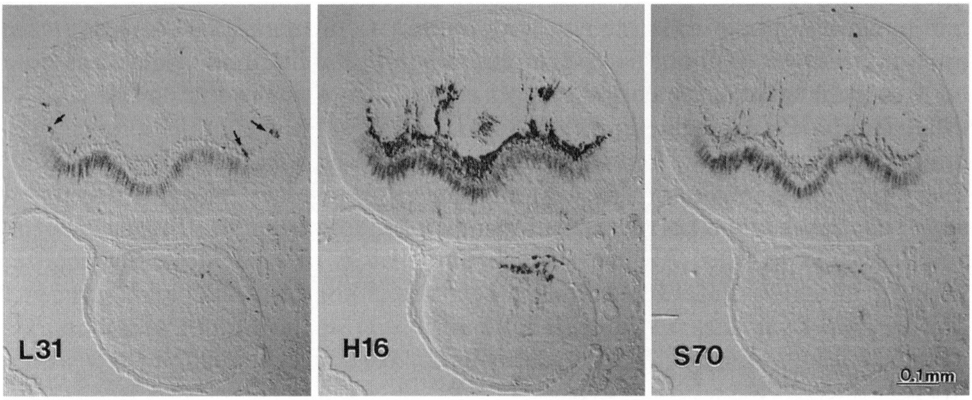

Fig. 4. Differential interference microscopy of serial trio out of transverse frozen section of lamprey pineal complex. The section contains middle portions of pineal end-vesicle (top) and parapineal vesicle (bottom). Note the clusters of melanin particles marking the luminal ends of pigment epithelial cells standing side by side with pineal photoreceptor cell. In the pineal end-vesicle, the outer segments project out of the floor into the lumen.

The central region is occupied by such outer segments (perhaps type 1) as strongly stained with DAB upon contact with S70, specific for rod-type short-cell outer segment, and with H16, more or less specific for both rod- and cone-type outer segments of the retina. In the periphery of the floor, a small number of cells react against L31, specific for cone-type long-cell outer segment of the retina. The roof plate of the pineal end-vesicle is pigment-free and known as "pellucida". In the parapineal floor, a very small number of cells react against each of the antibodies, L31, H16 or S70.

[11]. Regional differentiation of photoreceptor cells in the pineal floor will show a phase of specialization of the cells existing in a certain optimal location. Their central location will be regarded as an advantage for receiving radiation efficiently if it is given directly from above through the pellucida or pineal roof. Cells existing in the peripheral region of the floor presumably belong to the type 2 or cone type cells. Their morphology and immunoacitivity will be evidence of their transformation toward neuroendocrine cell lines, while retaining their inheritance of photoreception perhaps for only a selected range of light. Following the line of study, the authors are producing a series of antibodies, such as active for arrestin, Müller cell cytoplasm, horizontal cell cytoplasm, or pigment epithelial cytoplasm, etc. Each one and a combination of these monoclonal antibodies will form useful tools to explore the relationships among autologous and heterologous photoreceptors, related neuronal as well as glial systems.

Acknowledgements

This study has been supported by Grant-in-Aid, No. 09670001 to Akira Tonosaki from the Ministry of Education, Science and Culture, Japan. The authors are grateful to Prof M. Endo and Prof S. Yagihashi, Hirosaki University School of Medicine, for the generous invitation to the First Hirosaki International Forum of Medical Science, 30 October 1997. Appreciation is also addressed to Mr Isao Aihara in the Department for long-term technical assistance of study.

References

1. Schulze M. Sehorgan I. Die Retina. In: Stricker S (ed) Handbuch der Lehren von den Geweben, vol 2, Leipzig: Wilhelm Engelman, 1873;977–1034.
2. Bach L. Augengrube, Augenblase. In: Bach L, Seefelder R (eds) Atlas zur Entwicklungsgeschichte des Menschlichen Auges. Leipzig: Wilhelm Engelmann, 1911;5–12.
3. Frisch K v. Beiträge zur Physiologie der Pigmentzellen in der Fischhaut. Pflügers Arch 1911; 138:319–387.
4. Scharrer E. Die Lichtempfindlichkeit blinder Elritzen. I. Untersuchungen über das Zwischenhirn der Fische. Z Vegl Physiol 1929;7:1–38.
5. Oksche A. Historical perspectives of photoneuroendocrine systems. In: O'Brien PJ, Klein DC (eds) Pineal and Retinal Relationships. London: Academic, 1986;1–13.
6. García-Fernández JM, Foster RG. Immunohistochemical identification of photoreceptor proteins in hypothalamic cerebrospinal fluid-contacting neurons of the larval lamprey (*Petromyzon marinus*). Cell Tissue Res 1994;257:319–326.
7. García-Fernández JM, Jiménez AJ, González B, Pombal RG, Foster RG. An immunohistochemical study of encephalic photoreceptors in three species of lamprey. Cell Tissue Res 1997; 288:367–378.
8. Eakin RM. Continuity and diversity in photoreceptor cells. In: Eakin RM, Westfall JA (eds) Visual Cells in Evolution. New York: Paren, 1982;91–105.
9. Katori Y, Takasaka T, Ishikawa M, Tonosaki A. Fine structure and lectin histochemistry of the apical surface of the free neuromast of *Lampetra japonica*. Cell Tissue Res 1994;276:245–252.
10. Kikuchi T, Tonosaki A. Development of apical-surface structures of mouse optic placode. Acta Otolaryngol (Stockh) 1988;106:200–207.
11. Okano T, Kojima D, Fukuda Y, Schichida Y, Yoshizawa T. Primary structures of chicken cone visual pigments: vertebrate rhodopsins have evolved out of cone visual pigments. Proc Natl Acad Sci USA 1992;89:5932–5936.
12. Ishikawa M, Takao M, Washioka H, Tokunaga F, Watanabe H, Tonosaki A. Demonstration of rod and cone photoreceptors in the lamprey retina by freeze-replication and immunofluorescence. Cell Tissue Res 1987;256:227–232.
13. Negishi K, Teranishi T, Kuo CH, Miki N. Two types of lamprey retina photoreceptors immunoactive to rod- or cone-specific antibodies. Vision Res 1987;27:1237–1241.
14. Tonosaki A, Washioka H, Hara M, Ishikawa M, Watanabe H. Gap junctions and synaptic relations of horizontal cells in lamprey retina. Neurosci Res 1986;Suppl 6:S107–S118.
15. Tonosaki A, Yoshida M, Hara M, Ishikawa M, Watanabe H. Photoreceptor disk membrane of Lampetra japonica. Neurosci Res 1989;6:340–349.
16. Yoshida M, Saito S, Koike Y, Ishikawa M, Watanabe H, Tokunaga F, Tonosaki A. Anti-lamprey retinal antibodies: immunohisotochemistry on the retinas of several species of vertebrates. Cell Tissue Res 1991;266:419–426.
17. Yoshida M, Tonosaki A. Anti-lamprey (*Lampetra japonica*) retinal antibodies which distinguish between short and long photoreceptor cells. Comp Biochem Physiol 1994;107B:183–188.

Dynamic Cells: Cell Biology of the 21st Century
S. Yagihashi, T. Kachi and M. Wakui, editors.

Neurotransmission in the inner ear — glutamatergic transmission in the afferent synapses of hair cells

Shin-ichi Usami
Department of Otorhinolaryngology, Hirosaki University School of Medicine, Hirosaki, Japan

Keywords: aspartate, glutamate, immunocytochemistry, receptor, transporter.

Introduction

In the past 10 years, various neuroactive substances have been determined to be neurotransmitters and/or neuromodulators in the inner ear by means of electrophysiological and immunocytochemical studies. Furthermore, recently, molecular biology has enabled the structure of various neurotransmitter receptors and transporters to be discerned showing that a complex of receptors as well as transporters are involved in neurotransmission in the inner ear. The present study focuses mainly on neurotransmission between hair cells and primary afferents. Our recent immunocytochemical findings concerning the localization of key molecules which are relevant to afferent transmission (receptors, transporters, and related enzymes) are emphasized in this report.

Materials and Methods

A biochemical approach does not allow for the examination of the molecules at the cellular and subcellular level. Postembedding electron microscopic immunocytochemistry has been utilized in order to fulfill the resolution required.

Animals

For this series of studies, Wistar rats, Hartley guinea pigs, and squirrel monkeys were used. Details are described elsewhere (see [1—6]). Briefly, after being deeply anesthetized with sodium pentobarbital (50 mg/kg) the animals were perfused with a mixture of glutaraldehyde and paraformaldehyde as described below. They were then immunocytochemically processed in the following manner.

Address for correspondence: S. Usami, Department of Otorhinolaryngology, Hirosaki University School of Medicine, 5 Zaifu-cho, Hirosaki 036-8562, Japan. Tel.: +81-172-39-5099. Fax: +81-172-39-5100. E-mail: usami@cc.hirosaki-u.ac.jp

Fixatives

1. 2.5% glutaraldehyde + 1% paraformaldehyde (in 0.1 M phosphate buffer pH 7.4) was used to fix amino acids including glutamate, glutamine, and aspartate. The perilymphatic perfusion method was employed for electron microscopic observation, in order to assure adequate and rapid fixation. The temporal bones were then postfixed (4 h, room temperature) in the same fixatives.
2. 0.1% glutaraldehyde + 4% paraformaldehyde (in 0.1 M phosphate buffer pH 7.4) was chosen for aspartate, glutamate receptors, and transporters.

Immunocytochemical procedures

1. Postembedding light and electron microscopic immunocytochemistry using Durcupan-embedded material: for amino acid immunocytochemistry, postembedding immunocytochemistry was used to visualize each amino acid in various cell types of the inner ear [1,2].
2. Postembedding electron microscopic immunocytochemistry using the freeze substitution method: standard procedures include osmium tetroxide treatment and dehydration at room temperature followed by embedding in an epoxy resin, but in freeze substitution the use of osmium is omitted, allowing dehydration, embedding and polymerization to be carried out at subzero temperatures. This facilitates visualization of protein antigens such as glutamate receptors/transporters and indications are that a significant increase in the sensitivity of amino acid immunocytochemistry is also provided. Lowicryl HM 20 embedded material was processed as described in detail previously [4–6].
3. Light microscopic immunocytochemistry was carried out using cryostat sections [3].

Antisera and specificity controls

A series of antisera related to glutamate metabolism used in this study has been previously characterized (see [1–6]). Adsorption control was performed by adding to the primary antisera. For postembedding immunocytochemistry, test sections containing amino acid-glutaraldehyde-brain protein conjugates were incorporated in each experiment and immunocytochemical selectivity monitored.

Results and Discussion

Presynaptic localization of glutamate

One of the criteria to be fulfilled for a neurotransmitter is presynaptic localization. Postembedding immunocytochemistry has shown the cellular and subcellular localization of glutamate in the rat cochlea and vestibular endorgans [1,2]. In both the organ of Corti and vestibular endorgans, glutamate immunoreactivity was generally preferentially distributed in the hair cells indicating a compatibility for hair cell transmitters (Fig. 1). Through postembedding immunogold cyto-

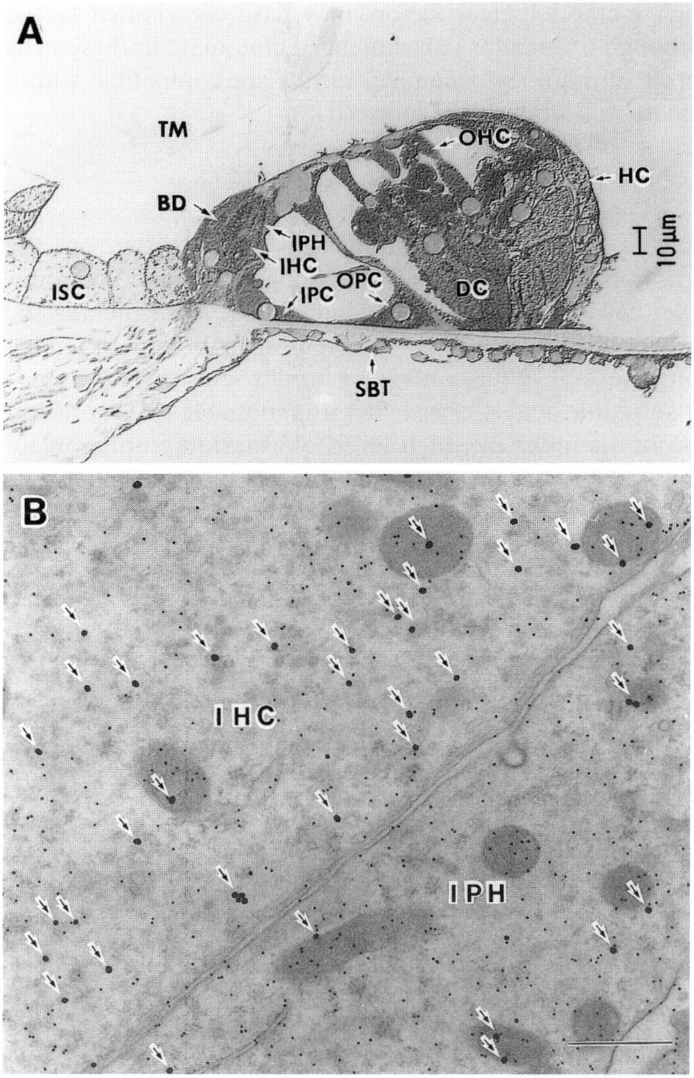

Fig. 1. Presynaptic localization of glutamate. **A:** Glutamate distribution in the guinea pig organ of Corti. IHCs and OHCs are heavily stained. Adjacent supporting cells, including border cells (BD), inner phalangeal cells (IPH), inner and outer pillar cells (IPC, OPC), Deiters cells (DC), and subbasilar tympanic cells (SBT) are also immunoreacted. Hensen cell (HC) and inner sulcus cells (ISC) are weakly stained. TM: tectorial membrane. Scale bar = 10 µm. **B:** Double-labelled ultrathin sections (glutamate: 30-nm gold particles indicated by arrows, glutamine: 15 nm) showing the subcellular localization of glutamate and glutamine in the rat organ of Corti. The density of gold particles signalling glutamate is higher in the IHC than in the adjacent supporting cell (IPH). Scale bar = 0.5 µm.

chemistry, it has been shown that gold particles representing glutamate are highly correlated with synaptic vesicles in the glutamatergic synapse in the CNS [7]. This has not been shown in hair cells in the inner ear, however [1,2]. Therefore, hair cells in the inner ear may possibly have a different release mechanism from those in the CNS, though methodological factors may have contributed to the differences observed. Although a vesicular enrichment of glutamate in these cells has not yet been shown, our immunocytochemical results are compatible with a transmitter role of glutamate in both types of hair cell.

Aspartate in hair cells

Although glutamate's role as a hair cell transmitter has been focused on, the identity of the excitatory amino acid that serves as an endogenous ligand of glutamate receptors has not yet been confirmed. Candidates in the CNS include the endogenous compounds with affinity to glutamate receptors including glutamate and aspartate, in addition to several sulfur-containing amino acids such as homocysteic acid and cysteine sulphinic acid. Consistent with a possible role for aspartate in neurotransmission in the inner ear, high levels of aspartate immunoreactivity were found in neuronal compartments, including all types of hair cells, and afferent and efferent nerve fibers (Fig. 2) [4]. However, a possible metabolic role for aspartate was also indicated by the strong immunoreactivity in subbasilar tympanic cells, interdental cells, and the mesothelial cells in the Reissner's membrane [4].

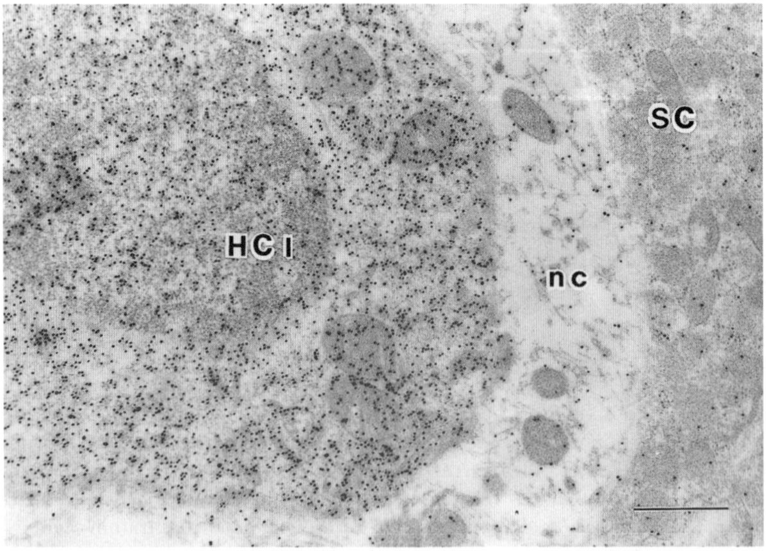

Fig. 2. Aspartate immunoreactivity in the rat vestibular endorgans. Aspartate is concentrated within the HCI, while the SC is much less intensely labelled. nc: nerve chalice. Scale bar = 0.5 μm.

Glutamate-glutamine cycle in the inner ear

Glutamine has previously been shown to be a major precursor and breakdown product of glutamate. It is thought that a glutamate-glutamine cycle is a critical feature of glutamatergic transmission in the CNS [8]. Glutamate is taken up by glial cells and converted to glutamine by the glial enzyme glutamine synthetase. Glutamine is transferred back to the neurons and converted to glutamate by phosphate-activated glutaminase, an enzyme assumed to reside primarily in neurons, completing the cycle. It is thought that this cycle assists in rapid removal of glutamate after release, avoiding interference with synaptic transmission and toxic buildup of glutamate. Immunocytochemical data have demonstrated that glutamate immunoreactivity is preferentially distributed in the hair cells, whereas glutamine immunoreactivity is enriched in supporting cells (SC) (Fig. 3) [1,2]. The glutamate:glutamine ratio varies according to cell type, but the ratio is statistically higher in hair cells, in contrast with various types of SCs. The metabolic compartmentation is similar to that of glutamatergic nerve terminals and adjacent glial processes in the central nervous system. Glutamate and glutamine synthetize enzymes and phosphate-activated glutaminase and glutamine synthetase maintain this gradience.

Immunocytochemical studies have shown a complementary localization pattern of glutamate synthetize enzyme being enriched in hair cells as opposed to the predominant localization of glutamine synthetize enzyme in the SCs [6,9]. This is consistent with the existence of a glutamate-glutamine cycle between

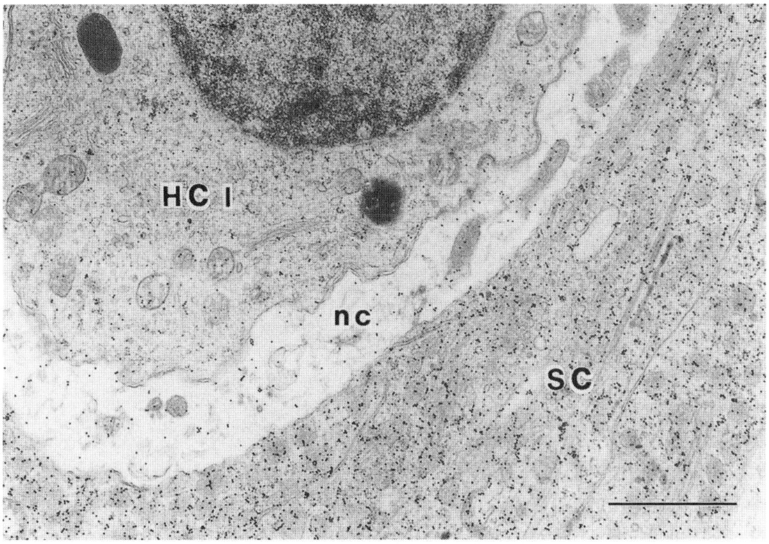

Fig. 3. Glutamine immunoreactivity in the rat vestibular endorgans. In contrast to glutamate/aspartate, glutamine is preferentially distributed in the SC. Scale bar = 1.0 μm.

hair cells and SCs, resembling the relationship between neuron and glial cells in the CNS.

Glutamate receptors in the inner ear

The type of glutamate receptor involved determines the cellular response to glutamate, therefore, much attention has been paid to the diversity of the glutamate receptors. It has been suggested by recent studies involving molecular techniques and antibodies specific for different receptor subunits, that various types of glutamate receptors, including AMPA, Kainate, and NMDA receptor subunits are expressed in the spiral and vestibular ganglia [10—14]. Metabotropic receptors have also been reported to contribute to hair cell/afferent neurotransmission in the inner ear [15]. The view that glutamate may serve as a hair cell transmitter has been further supported by our demonstration of immunocytochemical distribution of NMDA receptor subunits NMDAR1 and AMPA-selective glutamate receptor subunits (GluR1, GluR2/3, GluR4) in the spiral and vestibular ganglion (Fig. 4A). We also attempted to determine whether these glutamate receptor subunits additionally occur on the postsynaptic membrane. Concentration of GluR2/3 and 4 of the AMPA type glutamate receptors at the postsynaptic specialization of afferent synapses established by inner hair cells (IHC) in the organ of Corti was shown by high-resolution immunogold analysis (Fig. 4B,C) [5]. Immunoreactivities for GluR2/3 and GluR4 in the vestibular sensory epithelium were concentrated in the synapses between type I hair cells (HCI) and the surrounding chaliceal nerve endings and with the bouton type nerve endings contacting type II hair cells (HCII) (Fig. 4B). This suggests that AMPA receptor subunits have an involvement in the afferent transmission. A lack of detectable AMPA receptor immunoreactivity in the synaptic region of the outer hair cells (OHC) indicates considerable differences between inner and OHCs in regard to glutamatergic transmission. Regarding NMDA receptors, we were unable to localize them on the postsynaptic membrane at the ultrastructural level (unpublished data), despite significant mRNA signalling as well as immunolabeling in the ganglion cell cytoplasm [10,12,13]. It is still an open question as to how transmission is carried out through NMDA receptors. They may contribute to synaptic plasticity which usually occurs during development as well as during nerve injury recovery periods.

Glutamate transporters in the inner ear

Glutamate action must be finalized by the glutamate transporter. GLAST, which is a glial type glutamate transporter, and glutamine synthetase colocalized in SCs apposed to the immunonegative hair cells, were shown by confocal microscopy [6]. That GLAST was heterogeneously distributed along the plasma membranes of the SCs, with higher concentrations basally (at the level of the afferent synapses) than apically, was confirmed by immunoelectron microscopy (Fig. 5)

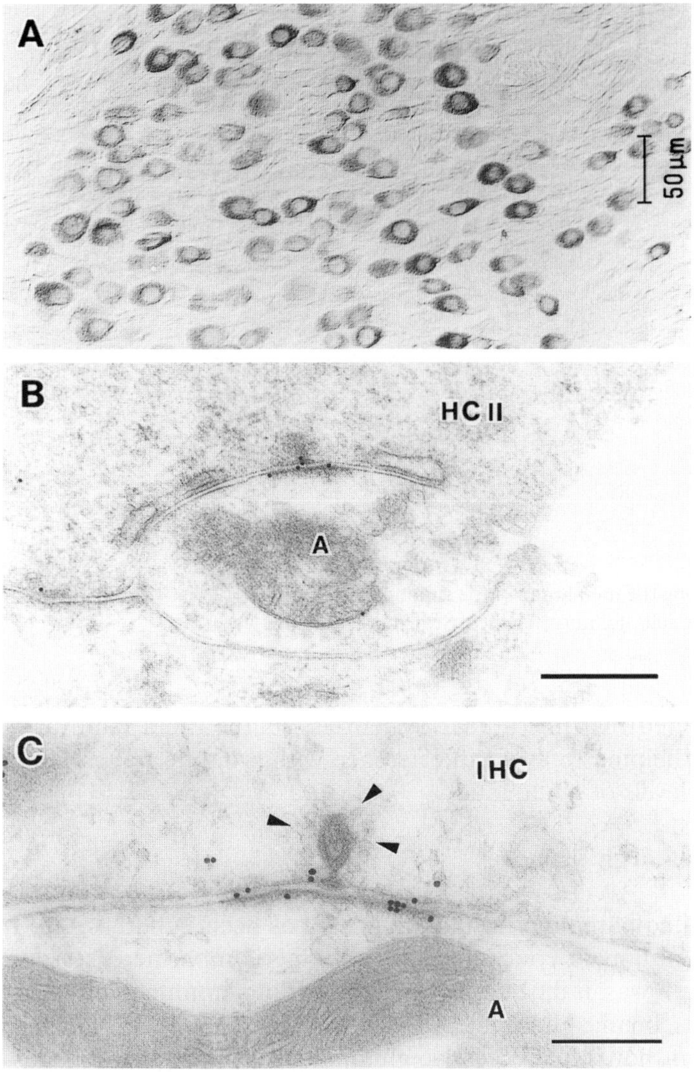

Fig. 4. AMPA type glutamate receptors at the hair cell/afferent synapses. **A:** Immunocytochemical localization of GluR2/3 receptor protein in the squirrel monkey vestibular ganglion. Cytoplasm of the ganglion cells are intensely immunoreactive. Scale bar = 50 µm. **B:** Gold particles signalling GluR4 were localized to the postsynaptic specialization of the afferent synapse (A) of the IHC. Arrowheads: synaptic body surrounded by synaptic vesicles. Some gold particles appear to be localized presynaptically. Scale bar = 0.3 µm. **C:** GluR2/3 is localized on the postsynaptic membrane of the afferent dendrite (A) in the rat vestibular endorgan. Scale bar = 0.5 µm. **C** is from [5].

[6]. It is suggested by the current findings that glutamate released at the afferent synapse of vestibular hair cells may be taken up by adjacent SCs and converted into glutamine. The SCs may, therefore, carry out at this peripheral synapse functions resembling those of glial cells in the CNS. GLAST was shown to be distrib-

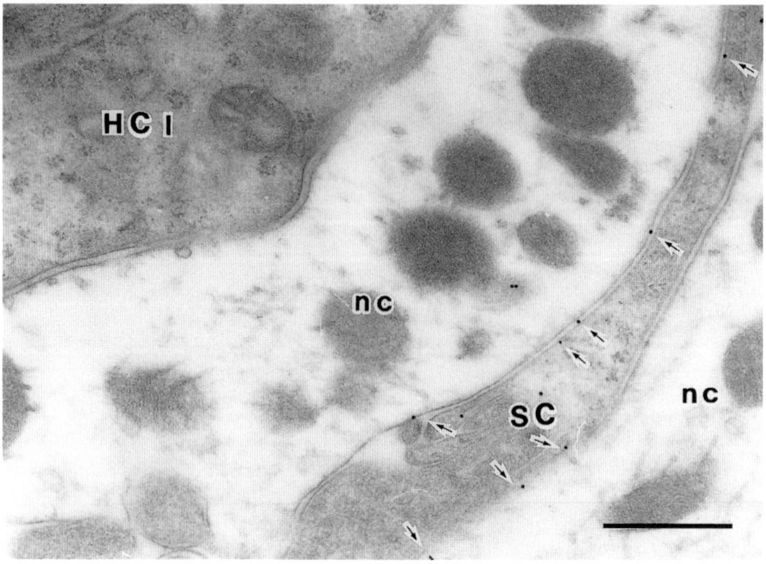

Fig. 5. GLAST is localized along the membrane of the supporting cell (SC) in the rat vestibular endorgan (arrows). HCI: type I hair cell; nc: nerve chalice. Scale bar = 0.5 μm.

uted in SCs. It is presumed that this uptake system assists the rapid removal of glutamate after release, helping to avoid interference with synaptic transmission and a build-up of toxic levels of glutamate.

Pathology of the glutamatargic synapses

Recently, glutamate-related pathology in the inner ear has been hypothesized. It has been reported that glutamate concentration in the perilymph increases during ischemia [16], and it was indicated by postembedding immunogold cytochemistry that during ischemia glutamate may be released from the hair cells and SCs [17]. It has been shown that excessive glutamate is exitotoxic and is possibly involved in ischemic neuronal damage in the CNS [18]. In agreement with these observations in the CNS, neurotoxic effects of glutamate have also been described in the inner ear. Swelling of afferent dendrites contacting IHCs has been induced by intracochlear perfusion of glutamate or glutamate agonist [19–21].

Conclusions

Although some controversy remains, the synapse between sensory hair cells in the inner ear and the afferent dendrites has been characterized, and now established as a model of glutamatergic synapse. This synapse is equipped with the key molecules which are relevant to glutamate transmission (receptors, transpor-

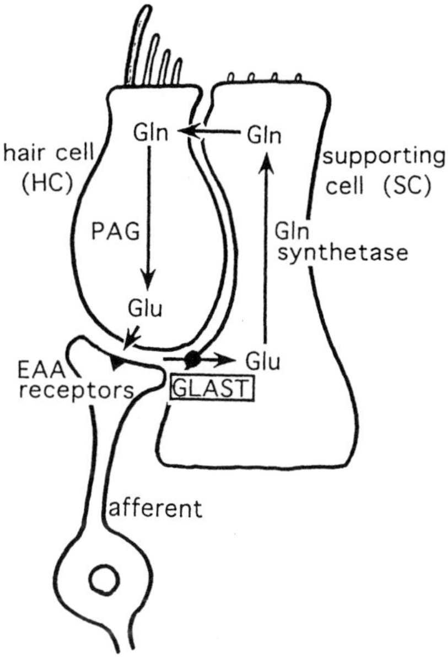

Fig. 6. Glutamate/glutamine cycle in the inner ear. Glutamate is concentrated in the hair cells and released. Afferent dendrites are equipped with several glutamate receptors on the postsynaptic membrane. Released glutamate escapes from the synaptic cleft by glutamate transporters, such as GLAST, into adjacent SCs. SCs in the inner ear may transport glutamate across the membrane and convert it to glutamine in the cytoplasm. Glutamate may be used again by hair cells as a neurotransmitter. This cycle between hair cells and SCs resembles the glutamate/glutamine cycle in the CNS.

ters, and related enzymes) (Fig. 6). Comparison with the glutamatergic synapse in the CNS revealed many similarities as well as many differences existing in this structurally and evolutionally distinct synapse. Whether perturbation of glutamatergic synapse could be of pathogenetic relevance in various inner ear diseases should now be explored.

Acknowledgements

This study was in done in collaboration with Dr Ole Petter Ottersen of the University of Oslo and Drs Atsushi Matsubara, Shigetoshi Fujita and Yutaka Takumi of Hirosaki University. This work was supported by a Grant-in-Aid for Scientific Research from the Ministry of Education, Science and Culture of Japan, the Ministry of Health and Welfare of Japan, and The Karohji Memorial Fund for Medical Research in Hirosaki University.

References

1. Usami S, Osen KK, Zhang N, Ottersen OP. Distribution of glutamate-like and glutamine-like immunoreactivities in the rat organ of Corti: a light microscopic and semiquantitative electron

microscopic analysis with a note on the localization of aspartate. Exp Brain Res 1992;91:1–11.

2. Usami S, Ottersen OP. Differential cellular distribution of glutamate and glutamine in the rat vestibular endorgans: an immunocytochemical study. Brain Res 1995;676:285–292.
3. Usami S, Matsubara A, Fujita S, Shinkawa Hayashi M. NMDA (NMDAR1) and AMPA-type (GluR2/3) receptor subunits are expressed in the inner ear. Neuro Report 1995;6:1161–1164.
4. Usami S, Ottersen OP. Aspartate is enriched in sensory cells and fibers in the guinea pig inner ear: a quantitative immunoelectron microscopic analysis. Brain Res 1996;742:43–49.
5. Matsubara A, Laake JH, Davanger S, Usami S, Ottersen OP. Organization of AMPA receptor subunits at a glutamate synapse: a quantitative immunogold analysis of hair cell synapses in the inner ear. J Neurosci 1996;16:4457–4467.
6. Takumi Y, Matsubara A, Danbolt NC, Laake JH, Storm-Mathisen J, Usami S, Shinkawa H, Ottersen OP. Discrete cellular and subcellular localization of glutamine synthetase and the glutamate transporter GLAST in the rat vestibular endorgan. Neuroscience 1997;(In press).
7. Ji Z, Aas JE, Laake J, Walberg F, Ottersen OP. An electron microscopic immunogold analysis of glutamate and glutamine in terminals of rat spinocerebellar fibers. J Comput Neurol 1991; 307:296–310.
8. Fonnum F. Glutamate: a transmitter in mammalian brain. J Neurochem 1984;42:1–11.
9. Ottersen OP, Takumi Y, Matsubara A, Landsend AS, Laake JH, Usami S. Molecular organization of a type of peripheral glutamate synapse: the afferent synapses of hair cells in the inner ear. Prog Neurobiol 1998;54:127–148.
10. Kuriyama H, Albin RL, Altschuler RA. Expression of NMDA-Receptor mRNA in the rat cochlea. Hear Res 1993;69:215–220.
11. Kuriyama H, Jenkins O, Altschuler RA. Immunocytochemical localization of AMPA selective glutamate receptor subunits in the rat cochlea. Hear Res 1994;80:233–240.
12. Fujita S, Usami S, Shinkawa H, Sato K, Kiyama H, Tohyama M. Expression of NMDA receptor subunit mRNA in the vestibular ganglion of the rat and guinea pig. Neuroreport 1994;5: 862–864.
13. Niedzielski AS, Wenthold RJ. Expression of AMPA, Kainate, and NMDA receptor subunits in cochlear and vestibular ganglia. J Neurosci 1995;15:2338–2353.
14. Demêmes D, Lleixa A, Dechesne CJ. Cellular and subcellular localization of AMPA-selective glutamate receptors in the mammalian peripheral vestibular system. Brain Res 1995;671: 83–94.
15. Safieddine S, Eybalin M. Expression of mGluR1α receptor mRNA in rat and guinea pig cochlear neurons. Neuroreport 1995;7:193–196.
16. Hakuba N, Gyo K, Yanagihara N, Mitani A, Kataoka K. Efflux of glutamate into the perilymph of the cochlea following transient ischemia in the gerbil. Neurosci Lett 1997;230:69–71.
17. Matsubara A, Kawabata Y, Takumi Y, Usami S, Shinkawa H, Haruta A, Matsuda K, Tono T. Quantitative immunogold cytochemistry reveals sources of glutamate release in inner ear ischemia. Acta Otolaryngol (Stockh) (Suppl.) 1998;(In press).
18. Rothman SM, Olney JW. Glutamate and the pathophysiology of hypoxic-ischemic brain damage. Ann Neurol 1986;19:105–111.
19. Pujol R, Rebillard G, Puel JL, Lenoir M, Eybalin M, Recasens M. Glutamate neurotoxicity in the cochlea: a possible consequence of ischemic or anoxic conditions occurring in ageing. Acta Otolaryngol (Stockh) (Suppl) 1991;476:32–36.
20. Pujol R, Puel J-L, D'aldin CG, Eybalin M. Pathophysiology of the glutamatergic synapses in the cochlea. Acta Otolaryngol (Stockh) 1993;113:330–334.
21. Puel JL, Pujol R, Tribillac F, Ladrech S, Eybalin M. Excitatory amino acid antagonists protect cochlear auditory neurons from excitotoxity. J Comput Neurol 1994;341:241–256.

Dynamic Cells: Cell Biology of the 21st Century
S. Yagihashi, T. Kachi and M. Wakui, editors.

Cornu Ammonis of carnivores: lack of a separate CA2 field

Kazuhiko Shoumura, Jin Hirama, Hideaki Amayasu, Sungman Yu, Hiroshi Yonekura and Noritaka Ichinohe
Department of Anatomy, Hirosaki University School of Medicine, Hirosaki, Japan

Abstract. Close comparison of Timm- and Nissl-stained adjoining sections has demonstrated the lack of a separate CA2 field in the cornu Ammonis of the cat and dog. In rodents, hares and primates, CA2 has been recognized as a discrete region which is characterized by large pyramidal cells like those of CA3, but does not receive mossy fiber inputs. In the cat and dog, however, mossy fiber projections extend all over the large pyramidal region of the cornu Ammonis. Furthermore, there are several focal mossy fiber projections in the proximal part of CA1, in which several, darkly stained, large pyramids, probably ectopic CA3 pyramidal cells, are intermingled with small CA1 pyramids. Thus, the cornu Ammonis of the cat and dog has no distinct field which matches CA2 in rodents, hares and primates.

Keywords: CA2, carnivores, cornu Ammonis.

Introduction

The essential structure of the mammalian hippocampal formation, comprised of the dentate gyrus, the cornu Ammonis or the hippocampus proper, and the subicular complex, is strikingly similar in widely divergent mammalian evolutionary radiations including monotremes, marsupials and uterine (placental) mammals. The cornu Ammonis and the dentate gyrus constitute the core of the hippocampal formation. The cornu Ammonis is subdivided into fields on the basis of cytoarchitectonic characteristics in Nissl-stained sections [1] or on the cytology and connectivity as revealed in Golgi-stained sections [2]. The nomenclature and subdivision of the cornu Ammonis introduced by Lorente de Nó [2] are most widely used, because the difference in the connectivity of each field is taken into consideration. According to this author, the cornu Ammonis is subdivided into four tangential fields: CA1, CA2, CA3 and CA4. Each of these fields is distinguished in a variety of mammals including mice, rats, monkeys and humans [2–5]. CA1–CA4 are also distinguished in armadillos, an edentate [6]. These findings led to a general remark that the basic structural organization of the cornu Ammonis is constant in mammalian phylogeny, although there are considerable differences in the cytoarchitectonic and connectional details of each field among different mammalian species [5]. However, recent studies from this laboratory ([7] and unpublished observations) have demonstrated the lack of

Address for correspondence: Kazuhiko Shoumura, Department of Anatomy, Hirosaki University School of Medicine, Hirosaki 036-8562, Japan.

a separate CA2 field in the cornu Ammonis of carnivores (cats and dogs). Comparison of the cornu Ammonis between carnivores and other mammals which have a distinct CA2 field is made in this paper.

Structure of the cornu Ammonis in rodents, hares, and primates with particular reference to CA2 field

In these mammalian species, CA1—CA4 are present as distinct fields of the cornu Ammonis. CA2 is the narrowest field of the cornu Ammonis interposed between CA3 and CA1. Large pyramidal cells of CA2 cannot be distinguished from those of CA3 in Nissl-stained sections, since pyramidal cell bodies of both fields are equally darkly stained. In Golgi sections, however, CA2 pyramids are distinguished from CA3 pyramids by the absence of thick thorns on the proximal apical dendrites that characterize CA3 pyramids [2,3,8]. Thick apical dendritic thorns of CA3 pyramids serve as the synaptic sites with giant terminals of mossy fibers emitted from the granular cells of the dentate gyrus.

CA2 can be also delineated from the neighboring CA3 and CA1 by comparing consecutive Timm- and Nissl-stained sections of the cornu Ammonis. Fig. 1A,B shows adjoining Timm- and Nissl-stained sections of the cornu Ammonis of the rat. Timm's stain [9] for zinc contained in high concentrations in the mossy fiber system [10] tapers to an end abruptly at the distal border of CA3 towards CA2 (Fig. 1A). The border between CA2 and CA1 can be readily identified in Nissl-stained sections by the sudden appearance of lightly stained, small pyramids of CA1 (Fig. 1B). The pyramidal cell layer of CA2 is bilaminated in primates, but is single layered in rodents (Fig. 1B) and hares (rabbits, unpublished observations). Although the pyramidal cell layer of CA3 is mostly composed of large pyramids, that of CA2 contains several, CA1-type, small pyramids intermingled with typical large pyramids of CA2 in rats [8] and rabbits (unpublished observations). In primates, these small pyramids are distributed chiefly in the deep pyramidal cell layer, while the superficial layer is composed mostly of large pyramids [5].

Chemical neuroanatomy has demonstrated several characteristic features of CA2 pyramids. In the human cornu Ammonis, CA2 is delineated from the neighboring CA1 and CA3 by immunohistochemical demonstration of calcium-binding proteins (calbindin-D28K) [11]; chromogranin A [12], which are especially rich in dentate granular cells and CA2 pyramids. In the rat, macaque monkey and baboon, however, most, but not all, of the pyramids of CA1 and CA2 are equally positive to calcium-binding protein immunoreactivity [13—15]. CA2 of human cornu Ammonis also stands out from neighboring CA3 and CA1 by its high content of lipofuscin pigment [3]. Furthermore, in the baboon, taurine-like immunoreactivity is selectively positive in the pyramids of CA2 and adjoining part of CA3 [16].

CA2 has some characteristic connectivity different from other fields of the cornu Ammonis. Axon collaterals of CA2 pyramids project to the neighboring parts

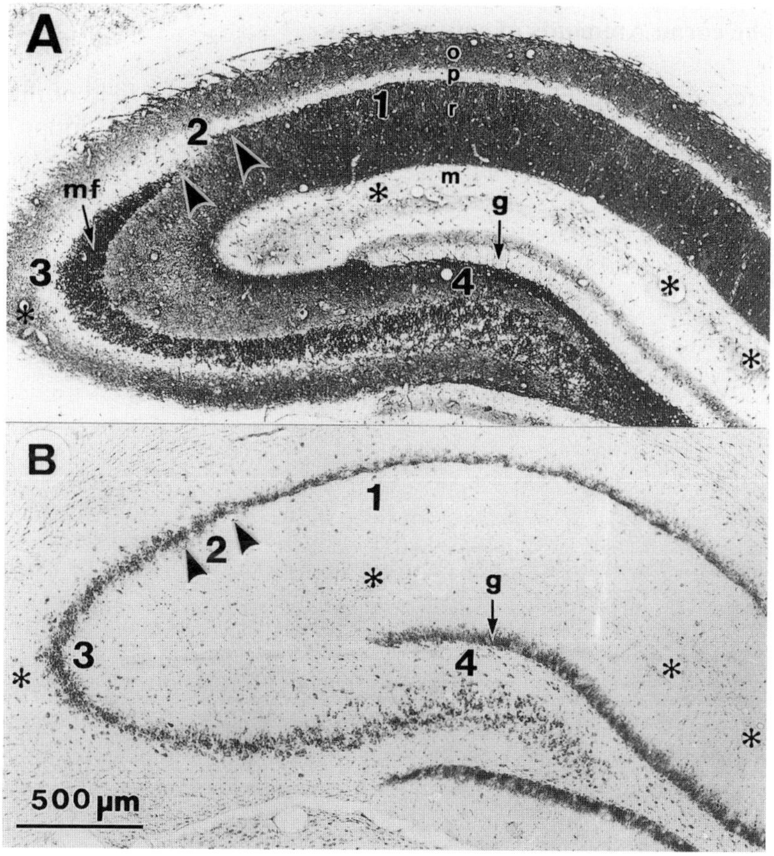

Fig. 1. Timm- (**A**) and Nissl-stained (**B**) adjoining transverse sections of the dorsal hippocampal formation of the rat. Abbreviations for Figs. 1—3: a, alveus; g, granular cells of the dentate gyrus; m, molecular layer; mf, mossy fiber layer; o, stratum oriens; p, stratum pyramidale; r, stratum radiatum; 1—4, CA1—CA4; D, dorsal; L, lateral; M, medial. In all figures, large arrowheads indicate the borders of the ammonic fields, and asterisks indicate identical blood vessels.

of CA1, CA3, and itself, and to the polymorphic cell layer of the dentate gyrus in the rat [17]. CA2 receives unique projections from the supramammillary nucleus of the hypothalamus in the monkey [18], although the projections are largely distributed in CA3 in the rat [19].

Ammon's horn sclerosis is a characteristic lesion of the temporal lobe epilepsies [20], which is illustrated by severe destruction of CA1 and CA3 pyramids, and sparing of CA2 pyramids and dentate granular cells. A unique population of neurons that survive in the sclerotic cornu Ammonis match those containing calcium-binding proteins [11,12], which are very likely to work as a buffer against seizure-induced calcium overload of the cytoplasm. There is, thus, a sufficient amount of anatomical and pathological evidence for considering CA2 as an entirely separate field from CA3 and CA1.

CA2 is absent in the cornu Ammonis of cats and dogs

Unlike rodents, hares, and primates, CA2 did not appear as a distinct field of the cornu Ammonis in cats and dogs. Figs. 2 and 3 show dorsal and ventral hippo-

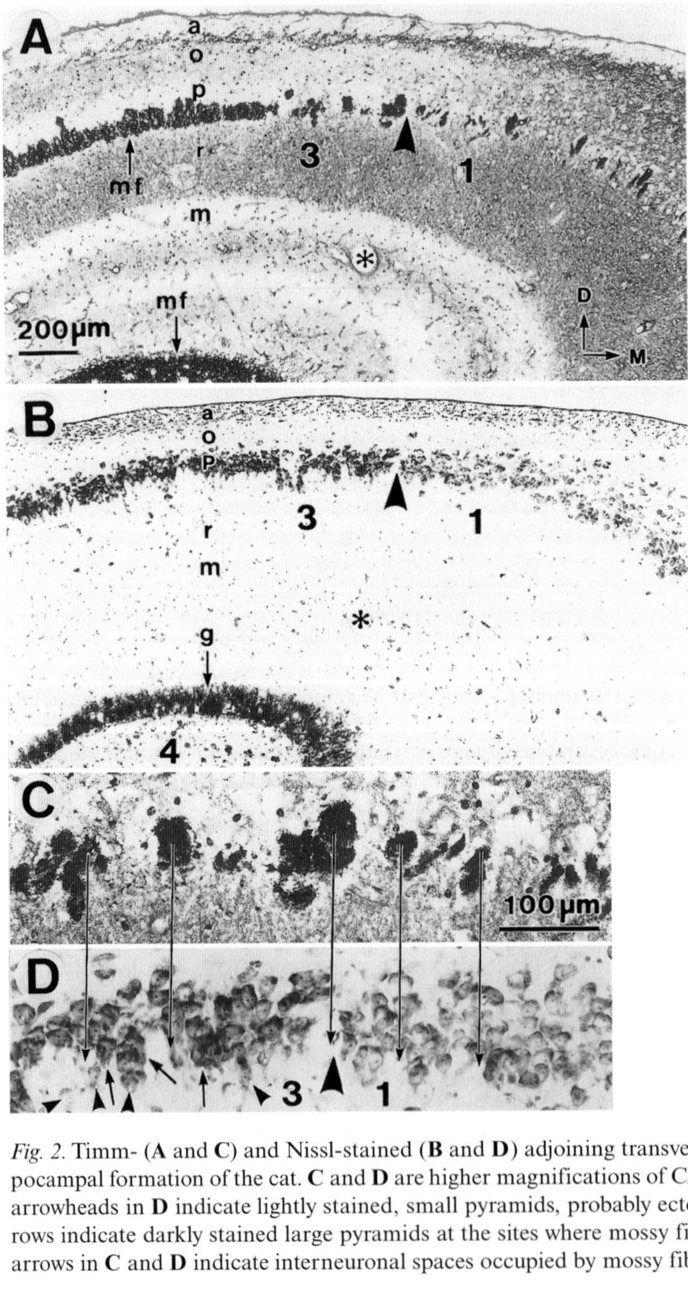

Fig. 2. Timm- (**A** and **C**) and Nissl-stained (**B** and **D**) adjoining transverse sections of the dorsal hippocampal formation of the cat. **C** and **D** are higher magnifications of CA3/CA1 border region. Small arrowheads in **D** indicate lightly stained, small pyramids, probably ectopic CA1 pyramids. Small arrows indicate darkly stained large pyramids at the sites where mossy fiber layer is interrupted. Long arrows in **C** and **D** indicate interneuronal spaces occupied by mossy fiber terminals.

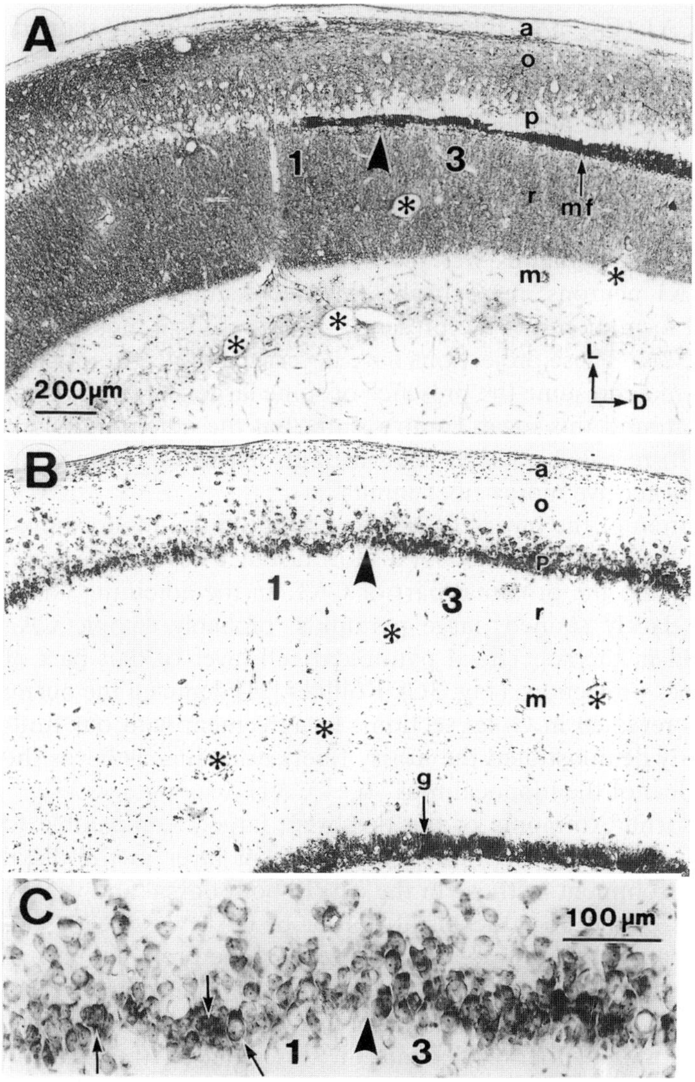

Fig. 3. Timm- (**A**) and Nissl-stained (**B** and **C**) adjoining transverse sections of the ventral hippocampal formation of the cat. C is CA3/CA1 border region at a higher magnification. Short arrows indicate darkly stained, large pyramids, probably ectopic CA3 pyramids, in the proximal part of CA1 which receives mossy fiber projections.

campal formations of the cat in adjoining Nissl- and Timm-stained [9] sections. The mossy fiber system stained by Timm's method filled the hilus of the dentate gyrus (polymorphic cells and CA4), extended partly through the stratum pyramidale at the proximal part of CA3, and continued throughout the stratum lucidum (i.e., mossy fiber layer) of the large-pyramidal region of the cornu Ammonis as far as the border with CA1 (Fig. 2A); the sudden appearance of lightly stained, small pyramids of CA1 in the Nissl-stained adjoining section allowed ready identification of the border (Fig. 2B,D). In the dorsal hippocampal formation, there were several small interruptions of mossy fiber layer in the distal part of CA3 (Fig. 2A,C). The adjoining Nissl-stained sections revealed a small number of darkly stained, large pyramids and lightly stained, small pyramids in the interrupted regions (Fig. 2D). From the morphology and the absence of mossy fiber inputs, these large and small pyramids were considered to correspond to CA2 neurons and ectopic CA1 neurons, respectively. Another distinctive finding in the dorsal hippocampal formation was the presence of several focal mossy fiber projections in CA1 (Fig. 2A). These projections have never been noted in rodents, hares, or primates. One may presume the presence of some large, darkly stained, ectopic CA3 pyramids around this focal Timm's stain, but the adjoining Nissl-stained sections (Fig. 2B) revealed no such neurons. This led us to assume that some CA1 neurons might receive mossy fiber inputs.

There was no interruption of the mossy fiber layer in the ventral hippocampal formation (Fig. 3A), but the layer continued for a short distance beyond the distal border of CA3 (Fig. 3B) into the proximal part of CA1. In the adjoining Nissl-stained section, a few, darkly stained, large pyramids, probably ectopic CA3 pyramids, were observed in the superficial pyramidal cell layer of that part of CA1 receiving mossy fiber projections (Fig. 3C). Kölliker [21] depicted the cornu Ammonis of a kitten as revealed in Golgi sections. In agreement with our findings, his figure 778 clearly demonstrated the mossy fibers extending as far as the border with the "dorsal leaf of the hippocampus" (i.e., CA1).

The findings in the cornu Ammonis of the dog were largely comparable to those in the cat. However, the interruptions of mossy fiber layer in the distal part of CA3 was not as prominent as those in the cat. In both dorsal and ventral hippocampal formations of the dog, the distal border of CA3 was clearly delineated like that in the cat, but there was a narrow zone in the proximal part of CA1 in which several, darkly stained, large pyramids were intermingled with lightly stained, small CA1 pyramids. In the adjoining Timm-stained sections, there were several focal mossy fiber projections in this narrow "transitional" zone. There was, again, no distinct region which matched Lorente de Nó's CA2 field in the cornu Ammonis of the dog.

Holm and Geneser [22] studied the cornu Ammonis of the domestic pig, by the modified Timm method. They subdivided the cornu Ammonis into regio superior and regio inferior [23], which are identical with CA1 and CA3/CA2, respectively. They stated that the main suprapyramidal part of mossy fibers extended as far as the limit towards the regio superior, and the infrapyramidal part had an extent reaching almost half-way to the regio superior. These findings strongly suggest that there is no region which matches Lorente de Nó's CA2 in the regio inferior of the cornu Ammonis of the pig like that of cats and dogs.

"Carnivores' cornu Ammonis" can be shared by several mammalian orders in the cohort Ferungulata

Phylogenetically, carnivores have been thought to be closer to ungulates than to other mammals [24]. A closer affinity between carnivores and artiodactyls (even-toed ungulates) than between the former and either rodents and primates has been suggested from comparison of DNA sequences of protein-coding genes [25]. We have demonstrated the lack of a separate CA2 field in the cornu Ammonis ("carnivores' cornu Ammonis") of the cat and dog, which is in striking contrast with those of rodents, hares, and primates. The pig, an artiodactyl, which is closer to carnivores than to rodents and primates in phylogeny, appears to have a "carnivores' cornu Ammonis". One possible explanation for these findings is that carnivores and artiodactyls evolved from a common ancestor who had a "carnivore's cornu Ammonis". Since carnivores and ungulates are grouped within the cohort Ferungulata [24], it seems possible that a "carnivores' cornu Ammonis" is shared by a wide variety of mammals within the cohort.

References

1. Rose M. Der Allocortex bei Tier und Mensch. J Psychol Neurol (Leipzig) 1926;24:1–260.
2. Lorente de Nó R. Studies on the structure of the cerebral cortex, II. Continuation of the study of the ammonic system. J Psychol Neurol (Leipzig) 1934;46:114–177.
3. Braak H. On the structure of the human archicortex, I. The cornu ammonis, a Golgi and pigmentarchitectonic study. Cell Tis Res 1974;152:349–383.
4. Cassel MD, Brown MW. The distribution of Timm's stain in the nonsulphide-perfused human hippocampal formation. J Comput Neurol 1984;222:461–471.
5. Rosene DL, Van Hoesen G. The hippocampal formation of the primate brain. A review of some comparative aspects of cytoarchitecture and connections. In: Jones EG, Peters A (eds) Cerebral cortex: Further Aspects of Cortical Function, Including Hippocampus, vol 6. New York: Plenum Press, 1987;345–456.
6. Travis MNG. Telencephalon of an edentate. In: Crosby EC, Schnitzlein HN (eds) Comparative Correlative Neuroanatomy of the Vertebrate Telencephalon. New York: Macmillan, 1982;338–382.
7. Hirama J et al. Cornu Ammonis of the cat: lack of a separate field of CA2. J Brain Res 1997; 38:487–493.
8. Ishizuka N et al. A quantitative analysis of the dendritic organization of pyramidal cells in the rat hippocampus. J Comput Neurol 1995;362:17–45.
9. Danscher G, Zimmer J. An improved Timm sulfide silver method for light and electron microscopic localization of heavy metals in biological tissue. Histochemistry 1978;55:27–40.
10. Frederickson CJ et al. Cytoarchitectonic distribution of zinc in the hippocampus of man and the rat. Brain Res 1983;273:335–339.
11. Sloviter RS et al. Calcium-binding protein (Calbindin-D28K) and parvalbumin immunohistochemistry in the normal and epileptic human hippocampus. J Comput Neurol 1991;308: 381–396.
12. Munoz DG. The distribution of chromogranin A-like immunoreactivity in the human hippocampus coincides with the pattern of resistance to epilepsy-induced neuronal damage. Ann Neurol 1990;27:266–275.
13. Sloviter RS. Calcium-binding protein (Calbindin-D28K) and parvalbumin immunohistochem-

istry: localization in the rat hippocampus with specific reference to the selective vulnerability of hippocampal neurons to seizure activity. J Comput Neurol 1989;280:183–196.
14. Baimbridge KG, Miller JJ. Immunohistochemical localization of calcium-binding protein in the cerebellum, hippocampal formation and olfactory bulb of the rat. Brain Res 1982;245:223–229.
15. Seress et al. Parvalbumin- and calbindin D28K-immunoreactive neurons in the hippocampal formation of the macaque monkey. J Comput Neurol 1991;313:162–177.
16. Ottersen OP et al. Taurine in the hippocampal formation of the Senegalese baboon, *Papio papio*: an immunocytochemical study with an antiserum against conjugated taurine. Exp Brain Res 1985;59:457–462.
17. Ishizuka N et al. Organization of intrahippocampal projections originating from CA3 pyramidal cells in the rat. J Comput Neurol 1990;295:580–623.
18. Veazey RB et al. The morphology and connections of the posterior hypothalamus in the cynomolgus monkey (*Macaca fascicularis*). II. Efferent connections. J Comput Neurol 1982;207: 135–156.
19. Haglund L et al. The projection of the supramammillary nucleus to the hippocampal formation: an immunohistochemical and anterograde transport study with the lectin PHA-L in the rat. J Comput Neurol 1984;229:171–185.
20. Gloor P. The Temporal Lobe and Limbic System. New York, Oxford: Oxford University Press, 1997.
21. Kölliker A. Handbuch der Gewebelehre des Menschen 6th ed. Leipzig: Engelmann, 1896. Cited in J Comput Neurol 1982;207:135–156.
2. Holm IE, Geneser EA. Histochemical demonstration of zinc in the hippocampal region of the domestic pig: II. Subiculum and hippocampus. J Comput Neurol 1991;305:71–82.
23. Ramón y Cajal S. Histology of the Nervous System, vol II. Swanson N, Swanson LW (trans). New York: Oxford University Press, 1995.
24. Simpson GG. The principles of classification and a classification of mammals. Bull Am Mus Nat Hist 1945;85:1–350.
25. Li W-H et al. Molecular phylogeny of rodentia, lagomorpha, primates, artiodactyla, and carnivora and molecular clocks. Proc Natl Acad Sci USA 1990;87:6703–6707.

Dynamic Cells: Cell Biology of the 21st Century
S. Yagihashi, T. Kachi and M. Wakui, editors.

Gene expression and regulation of Ca^{2+}/calmodulin-dependent protein kinase type IV in the nervous tissues

Hiroyuki Sakagami and Hisatake Kondo
Division of Histology, Department of Cell Biology, Graduate School of Medical Sciences, Tohoku University, Sendai, Japan

Abstract. Ca^{2+}/calmodulin-dependent protein (CaM) kinase IV is a monomeric multifunctional enzyme and has two isoforms termed α and β. CaM kinase IV β is different from the α isoform due to the presence of a 28 amino acid N-terminal extension. Due to this difference alone, the gene expression for CaM kinase β is confined to the cerebellar granule cells, whereas that for CaM kinase α is distributed widely in the forebrain as well as the cerebellar granule cells. Concerning its subcellular localization, CaM kinase IV is predominantly localized in the neuronal cell nuclei in the central as well as peripheral nervous system. CaM kinase IV in the neurons is likely to act as a mediator of Ca^{2+}-dependent gene induction through phosphorylation of transcription factors such as the cAMP-response element-binding protein and serum response factor. CaM kinase IV requires phosphorylation for its activation by CaM kinase kinase (CaMKK) which consists of two isoforms termed α and β. From the localization patterns for their mRNAs in the brain, it is likely that CaMKK α may be responsible for the activation of CaM kinase I in vivo, whereas CaMKK β may be responsible for that of CaM kinase IV.

Keywords: Ca^{2+}/calmodulin-dependent protein kinase IV, immunohistochemistry, in situ hybridization, isoform, nucleus.

Introduction

The calcium ion (Ca^{2+}) is a well-known second messenger in neurons as well as in other cells. External signals, such as hormones and transmitters, and electrical activities control movement of Ca^{2+} across the plasma membrane into the cytosol, and another second messenger, inositol triphosphate (IP3) regulates movement of Ca^{2+} out of the internal storage pool. The resulting rapid increases in cytosolic Ca^{2+} concentration serve as regulatory signals in the neurons. Most of the intracellular effects of Ca^{2+} are mediated through binding to various high-affinity Ca^{2+}-binding proteins, among which a 150-kDa protein termed calmodulin is most important in the Ca^{2+}-signal transduction: as the cytosolic Ca^{2+} concentration rises to the μM level, calmodulin is changed into a multifunctional activator by occupancy of its four Ca^{2+}-binding domains with Ca^{2+} and it regulates a number of other proteins including several protein kinases.

Ca^{2+}/calmodulin-dependent protein (CaM) kinases represent a major family

Address for correspondence: Prof Hisatake Kondo, Division of Histology, Department of Cell Biology, Graduate School of Medical Sciences, Tohoku University, Sendai, Japan. Fax: +81-22-717-8035. E-mail: hkondo@mail.cc.tohoku.ac.jp of such target proteins, including CaM kinase I, II, III and IV, phosphorylase kinase, and myosin light chain kinases. Among them, CaM kinases I, II and IV have been classified into multifunctional CaM kinases, which can phosphorylate a variety of substrates [1–3]. CaM kinase II is an oligomeric enzyme composed of four subunits and is extremely abundant in the brain. The enrichment of CaM kinase II in the postsynaptic density as well as presynaptic terminals implies its functional significance in neuronal transmission, especially the involvement in memory, which has been extensively studied [1,3–5]. On the other hand, CaM kinases I and IV are monomeric enzymes. CaM kinase IV was identified relatively recently by screening a brain cDNA expression library with radio-labelled calmodulin, and with a monoclonal antibody against a calmodulin-sensitive epitope on CaM kinase II [6,7].

Molecular structure

Using Western blotting and SDS-PAGE gel electrophoresis of a purified native enzyme from the rat cerebellum, the existence of two polypeptides with molecular weight values of 63,000–65,000 and 66,000–67,000 was demonstrated for CaM kinase IV, and they were termed α and β isoforms [7]. Subsequent studies on cDNA cloning of rat CaM kinase IV have revealed that the β isoform has a deduced 502 amino acid sequence containing catalytic and calmodulin-binding domains, while the α isoform is shorter by 28 amino acids on its N terminal [8,9]. Thus, Met-29 in the β isoform is the initiating Met for the α isoform. Analysis of the rat CaM kinase gene has revealed that the two isoforms are generated from the single gene by alternative transcriptional initiation [10].

CaM kinase IV is composed of a catalytic domain containing an ATP-binding site on its N terminal side and a regulatory domain containing CaM-binding site on its C terminal side (Fig. 1). This domain composition is basically common to CaM kinases I and II. A phosphorylation site (Thr-196) of CaM kinase IV α by CaM kinase kinase (CaMKK) is located in subdomain VIII of the protein

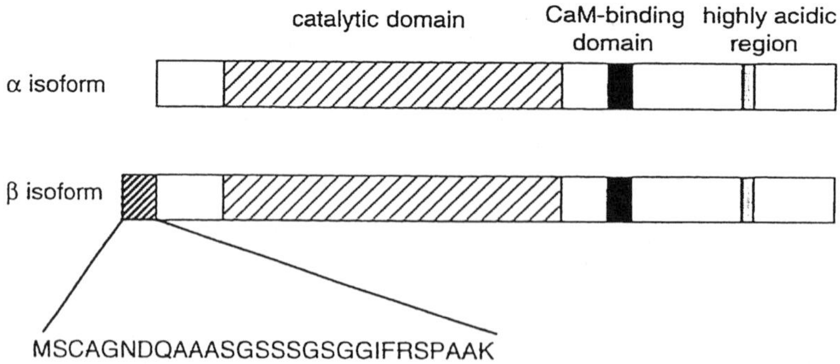

Fig. 1. Schematic representation of the primary structures of α and β isoforms of CaM kinase IV. The β isoform differs from the α isoform only by a unique 28 amino acid N-terminal extension.

kinases in a region known as the activation loop, resulting in the maximal activation of CaM kinase IV. The C terminal region next to the regulatory domain of CaM kinase IV contains a sequence of polyglutamic acids which is presumed to be responsible for interaction of this enzyme with intranuclear chromatins and consequent nuclear localization. The N terminal region distal to the ATP-binding site is longer than those of CaM kinases I and II and is rich in serine residues, which has been postulated to be autophosphorylated for relief of autoinhibition and acquisition of autonomous activity [11].

Localization and functional implication in the nervous system

In the Western blotting analysis using the antibody against CaM kinase IV, the two immunoreactive bands corresponding to the apparent molecular weights of the α and β isoforms are clearly detected only in the cerebellum, whereas a single band corresponding to the α isoform is detected in the forebrain [7]. With the Northern blot analysis, the hybridization signal is first detected in the rat brain on embryonic day 18 and it increased in intensity progressively thereafter, reaching a maximal level during the 2nd and 3rd postnatal week [12]. A recent study has demonstrated that CaM kinase IV is induced at transcriptional and translational levels specifically by the thyroid hormone 3,3′,5-triiode-thyronine in the primary cultured cells of the rat fetal telencephalon [13]. In addition, the 5′ flanking sequence of the β isoform of CaM kinase IV has revealed the presence of a putative thyroid hormone receptor core recognition motif. All these findings suggest that the expression of CaM kinase IV is under the control of the thyroid hormone.

When the localization of mRNA is examined with in situ hybridization histochemistry using oligonucleotide probes for 5′ noncoding nucleotide sequences specific to the α or β isoform, hybridization signals for the α isoform mRNA are deposited densely in the cerebellar granule cells, the olfactory granule cells, and the piriform cortex, and moderately in the cerebral neocortex, the hippocampal neuronal layers, and the caudate putamen, while those for the β isoform mRNA are detected solely in the cerebellar granule cells (Fig. 2) [9]. This indicates that promoters responsible for transcription of the α and β isoforms are spatially differentially regulated.

With immunohistochemistry, using the antibody against CaM kinase IV, the immunoreactivity appears most intensely in the cerebellar granule cells, and moderately in the cerebral neocortex, the piriform cortex, the hippocampal formation, the olfactory granule cells, the olfactory tubercle, the caudate putamen, and the thalamus in the adult rat brain (Fig. 3A,B) [7,14,15]. This immunoreactive pattern is basically consistent with the mRNA expression pattern described above. Weak immunoreactivity is also found in neurons in several brain stem nuclei such as the pontine nuclei, the locus ceruleus and the inferior olive, and the dorsal horn in the spinal cord. The immunoreactive products are predominantly localized in neuronal cell nuclei, although weak immunoreactivity is

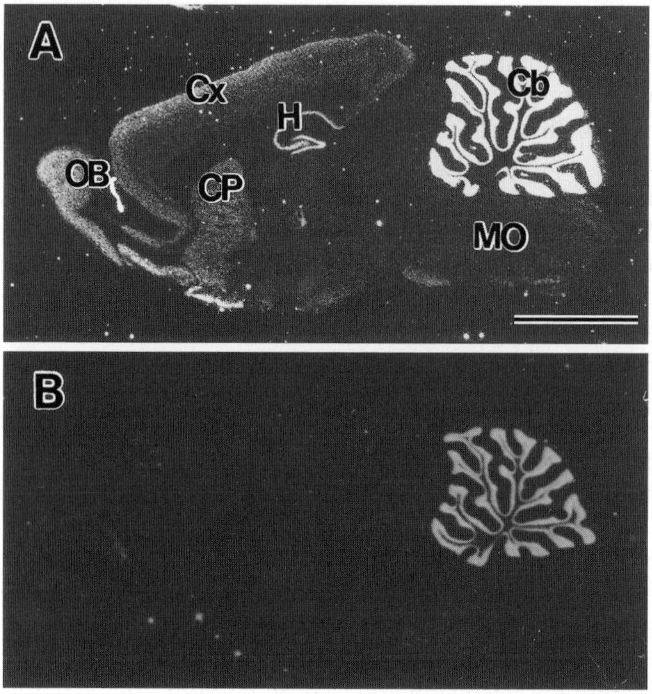

Fig. 2. Gene expression of CaM kinase IV α and β in the adult rat brain. Note the wide expression of CaM kinase IV α in contrast to restricted expression of CaM kinase IV β to the cerebellar granule cell layer. Cb: cerebellar cortex; CP: caudate putamen; Cx: cerebral cortex; H: hippocampal formation; MO: medulla oblongata; OB: olfactory bulb. Bar = 0.5 mm.

detected in their cytoplasm and dendritic processes (Fig. 3B) [7,14–16].

In the cerebellum, the immunoreactivity is observed in the granule cells and their axons in the molecular layer using antibodies which recognize both α and β isoforms of CaM kinase IV. We have recently succeeded in raising a polyclonal antibody against the β isoform using the synthetic peptide specific to the N terminal side of the β isoform. In accordance to its gene expression, the immunoreactivity for the β isoform is confined to the cerebellar granule cell layer. However, its immunoreactivity is confined to the cell nuclei of the granule cells and their axoplasm in the molecular layer does not exhibit any immunoreactivity at all. This finding confirms the β isoform as a distinct entity (Sakagami and Kondo, in preparation). Further study remains to be done in order to elucidate the functional difference between α and β isoform. Although cerebellar Purkinje cells in the adult brain do not exhibit CaM kinase IV at either mRNA or protein level, these neurons transiently express this molecule at late embryonic and early postnatal stages [12,17], which suggests some yet to be defined involvement of this kinase in the maturation of Purkinje cells in a time-dependent manner.

Outside the brain, the immunoreactivity is also detected in the sensory, autonomic and enteric neurons, and adrenal chromaffin cells [14]. The immunoreac-

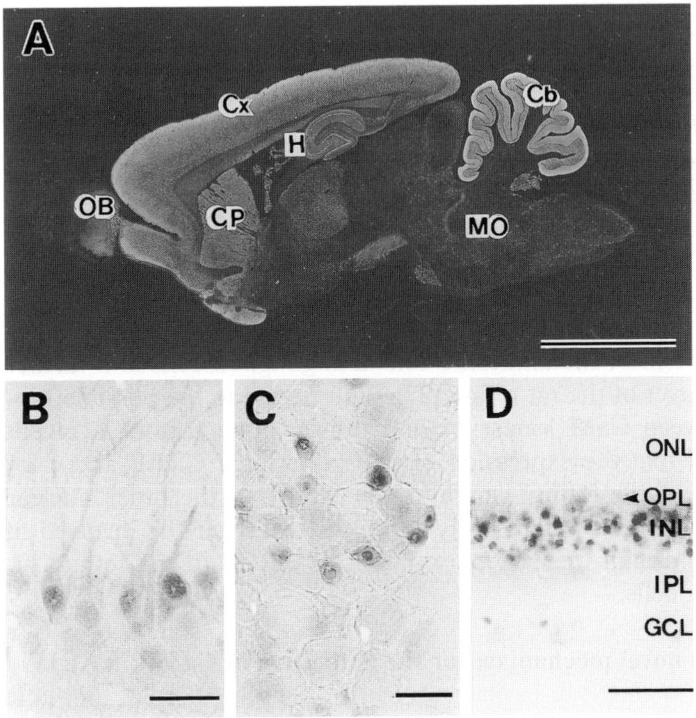

Fig. 3. Immunohistochemical localization of CaM kinase IV in the nervous system. **A:** Negative photographic image of CaM kinase IV-immunoreactivity in a sagittal section of adult rat brain. Note CaM kinase IV-immunoreactivity in a similar pattern to that of its gene expression. Cb: cerebellar cortex; CP: caudate putamen; Cx: cerebral cortex; H: hippocampal formation; MO: medulla oblongata; OB: olfactory bulb. Bar = 0.5 mm. **B:** Intense immunoreactivity in the nuclei of the hippocampal pyramidal cells. Bar = 50 μm. **C:** The immunoreactivity in small- to medium-sized neurons in the spinal ganglion. Bar = 50 μm. **D:** The immunoreactivity in the inner nuclear (INL) and ganglion cell (GCL) layers of the retina. IPL: inner plexiform layer; ONL: outer nuclear layer; OPL: outer plexiform layer. Bar = 100 μm.

tivity appears dominantly in the cell nuclei and immunoreactive nerve fibers are scarcely found as observed in central neurons. The nucleolus is free of immunoreactivity. In the spinal sensory neurons of adult rats, CaM kinase IV-immunoreactivity is present in about one-quarter of all calcitonin gene-related peptide (CGRP)-positive neurons and in the vast majority of the somatostatin-positive neurons (Fig. 3C) [18]. The immunoreactivity in the sensory neurons decreased markedly after axotomy [18].

The relatively restricted localization of CaM kinase IV in the nervous system and its intracellular localization confined to the nucleus are in contrast to those of CaM kinase I and II which are cytoplasmic and widely distributed in almost all neurons [19–21]. The nuclear localization strongly suggests the involvement of this enzyme in intranuclear Ca^{2+}-signal transduction. There has been evidence to show that transcription factors such as CREB (cAMP response element-binding protein), ATF-1 (activating transcription factor-1) and SRF (serum response factor) are among possible candidates for the intranuclear phosphoproteins [22–24]. It has been reported that CaM kinase IV is involved in the Ca^{2+}-dependent regulation of the expression of immediate early genes through phosphorylating CREB in cultured hippocampal neurons [25].

In the retina, only the α isoform of CaM kinase IV is detected at both protein and mRNA levels, and it is localized in most of the bipolar cells and some ganglion cells in which the immunoreactivity is largely confined to the cell nuclei like other immunoreactive neurons (Fig. 3D) [26]. This retinal localization is in contrast to that of CaM kinase II: CaM kinase II-immunoreactivity is localized in the perikaryal cytoplasm of the amacrine cells and ganglion cells and further in the inner plexiform layer of the rat retina [27]. With regard to the possible molecular interaction between CaM kinase IV and transcription factors, a recent study has demonstrated that the expression of phosphorylated CREB and c-fos increases in cell nuclei of the retinal ganglion cell layer and the inner nuclear layer after flashing and sustained light [28]. CaM kinase IV may be involved in the induction of c-fos through the phosphorylation of CREB in the retinal bipolar cells and ganglion cells.

CaM kinase cascade: a novel mechanism for the activation of CaM kinase IV

In contrast to CaM kinase II, whose activity is regulated through autophosphorylation by itself [1,2], CaM kinase IV, as well as CaM kinase I, is activated by novel upstream CaM kinases, CaMKKs [29,30]. Recent cDNA cloning analyses have revealed the deduced amino acid sequences of two CaMKKs termed α and β [31–33]: CaMKK α consists of 505 amino acids with a molecular weight of 55,907 while CaMKK β consists of 587 amino acids with a molecular weight of 64,445. The homologies between the two molecules in terms of the nucleotide sequences of the coding regions and the amino acid sequences are estimated to be approximately 65 and 69%, respectively.

In Western blotting analyses of adult rat brain using antibodies against synthetic peptides corresponding to the carboxyl terminal region of each CaMKK, the immunoreactivity for CaMKK α is detected densely in the cerebrum and brain stem, and weakly in the cerebellum and retina, whereas that for CaMKK β is detected in the cerebrum and cerebellum at a similar intensity and weakly in the brain stem. No significant immunoreactive band for CaMKK β is detected in the retina [32,33].

Using in situ hybridization histochemistry of adult rat brain, CaMKK α mRNA is localized widely in the gray matter throughout the entire brain, whereas CaMKK β mRNA is localized dominantly in the telencephalic and cerebellar gray matter (Fig. 4) [34]. When compared with the gene expression patterns in the adult brain for CaM kinase I [35] and CaM kinase IV [9,12], the pattern for CaMKK α is more similar to that for CaM kinase I, while that for CaMKK β is more similar to that for CaM kinase IV. These similarities suggest that

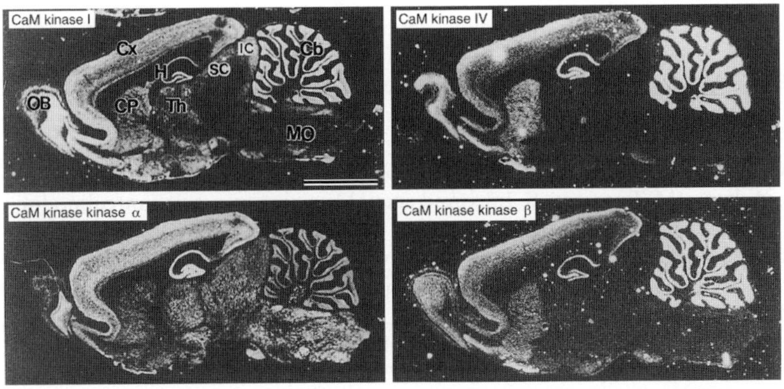

Fig. 4. Gene expression of the mRNAs for CaM kinase I, IV, CaM kinase kinase α and β in sagittal sections of adult rat brain. Note the good correlation in the expression pattern between CaM kinase I and CaM kinase kinase α and between CaM kinase IV and CaM kinase kinase β. Cb: cerebellar cortex; CP: caudate putamen; Cx: cerebral cortex; H: hippocampal formation; IC: inferior colliculus; MO: medulla oblongata; OB: olfactory bulb; SC: superior colliculus; Th: thalamus. Bar = 0.5 mm.

CaMKK α is more likely to be responsible for the activation of CaM kinase I, while CaMKK β is more likely to be responsible for that of CaM kinase IV in vivo. However, a recent immunohistochemical finding by Nakamura et al. [36] has demonstrated that the immunoreactivity for CaMKK α is localized predominantly in the neuronal cell nuclei, which is inconsistent with the cytoplasmic localization of CaM kinase I, but is rather consistent with the nuclear localization of CaM kinase IV. This discrepancy suggests that more complicated mechanisms for the activation of CaM kinase I and IV through two isoforms of CaMKK may exist in region-specific manners in vivo. Further studies on more details of biochemical characteristics as well as the cellular and subcellular localization of the two isoforms by immunohistochemistry will provide us with a key to understand more clearly the mechanism for the activation of CaM kinases.

Acknowledgements

This study was supported by Grant No. 08307001 and 09770002 to H. Sakagami and No. 09259203 and 09470001 to H. Kondo from the Ministry of Education, Science and Culture of Japan, and by a grant from Tanaka Orthopedic Clinic Foundation, Hamamatsu, Japan.

References

1. Hanson PI, Schulman H. Neuronal Ca^{2+}/calmodulin-dependent protein kinases. Ann Rev Biochem 1992;61:559–601.

2. Schulman H. The multifunctional Ca^{2+}/calmodulin-dependent protein kinases. Curr Biol 1993; 5:247–253.
3. Nairn AC, Picciotto MR. Calmodulin-dependent protein kinases. Cancer Biol 1994;5:295–303.
4. Silva AJ, Stevens CF, Tonegawa S, Wang Y. Deficient hippocampal long-term potentiation in α-calcium-calmodulin kinase II mutant mice. Science 1992;257:201–206.
5. Silva AJ, Payor R, Wehner JM, Tonegawa S. Impaired spatial learning in α-calcium-calmodulin kinase II mutant mice. Science 1992;257:206–211.
6. Sikela JM, Hahn WE. Screening an expression library with a ligand probe: isolation and sequence of a cDNA corresponding to a brain calmodulin-binding protein. Proc Natl Acad Sci USA 1987;84:3038–3042.
7. Ohmstede CA, Jensen KF, Sahyoun NE. Ca^{2+}/calmodulin-dependent protein kinase enriched in cerebellar granule cells. J Biol Chem 1989;264:5866–5875.
8. Means AR, Cruzalegui F, LeMagueresse B, Needleman DS, Slaughter GR, Ono T. A novel Ca^{2+}/calmodulin-dependent protein kinase and a male germ cell-specific calmodulin-binding protein are derived from the same gene. Molec Cell Biol 1991;11:3960–3971.
9. Sakagami H, Kondo H. Cloning of the β polypeptide of Ca^{2+}/calmodulin-dependent protein kinase IV and its expression confined to the mature cerebellar granule cells. Molec Brain Res 1993;19:215–218.
10. Sun Z, Means RL, LeMagueresse B, Means AR. Characterization and analysis of the complete rat calmodulin-dependent protein kinase IV gene. J Biol Chem 1995;279:29507–29514
11. Chatila T, Anderson KA, Ho N, Means AR. Unique phosphorylation-dependent mechanism for the activation of novel Ca^{2+}/calmodulin-dependent protein kinase type IV/Gr. J Biol Chem 1996;271:21542–21548.
12. Sakagami H, Watanabe M, Kondo H. Gene expression of Ca^{2+}/calmodulin-dependent protein kinase of the cerebellar granule cell type or type IV in the mature and developing rat brain. Molec Brain Res 1992;16:20–28.
13. Kreb J, Means RL, Honegger P. Induction of calmodulin kinase IV by the thyroid hormone during the development of rat brain. J Biol Chem 1996;271:11055–11058.
14. Sakagami H, Tsubochi H, Kondo H. Immunohistochemical localization of Ca^{2+}/calmodulin-dependent protein kinase IV in the peripheral ganglia and paraganglia of developing and mature rats. Brain Res 1994;666:173–181.
15. Nakamura Y, Okuno S, Sato F, Fujisawa H, An immunohistochemical study of Ca^{2+}/calmodulin-dependent protein kinase IV in the rat central nervous system: light and electron microscopic observations. Neuroscience 1995;68:181–194.
16. Jensen KF, Ohmstede CA, Fisher RS, Sahyoun N. Nuclear and axonal localization of Ca^{2+}/calmodulin-dependent protein kinase type Gr in rat cerebellar cortex. Proc Natl Acad Sci USA 1991;88:2850–2853.
17. Jensen KF, Ohmstede CA, Fisher RS, Olin JK, Sahyoun N. Acquisition or loss of a Ca^{2+}/calmodulin-dependent protein kinase during neuronal differentiation. Proc Natl Acad Sci USA 1991;88:4050–4053.
18. Ji RR, Shi TJ, Xu ZQ, Zhang Q, Sakagami H, Tsubochi H, Kondo H, Hokfelt T. Ca^{2+}/calmodulin-dependent protein kinase IV in dorsal root ganglion: colocalization with peptides, axonal transport and effect of axotomy. Brain Res 1996;721:167–173.
19. Ouimet CC, McGuinness TL, Greengard P. Immunohistochemical localization of calcium/calmodulin-dependent protein kinase II in rat brain. Proc Natl Acad Sci USA 1984;81:5604–5608.
20. Fukunaga K, Goto S, Miyamoto E. Immunohistochemical localization of Ca^{2+}/calmodulin-dependent protein kinase II in rat brain and various tissues. J Neurochem 1988;51:1070–1078.
21. Picciotto MR, Zoli M, Bertuzzi G, Nairn AC. Immunohistochemical localization of calcium/calmodulin-dependent protein kinase I. Synapse 1995;20:75–84.
22. Sun P, Enslen H, Myung PS, Maurer RA. Differential activation of CREB by Ca^{2+}/calmodulin-

dependent protein kinases type II and type IV involves phosphorylation of a site that negatively regulates activity. Genes Devel 1994;8:2527–2539.

23. Matthews RP, Guthrie CR, Wailes LM, Zhao X, Means AR, McKnight GS. Calcium/calmodulin-dependent protein kinase type II and IV differentially regulate CREB-dependent gene expression. Molec Cell Biol 1994;14:6107–6116.
24. Miranti CK, Ginty DD, Huang G, Chatila T, Greenberg ME. Calcium activates serum response factor-dependent transcription by a ras- and elk-1-independent mechanism that involves a Ca^{2+}/calmodulin-dependent kinase. Molec Cell Biol 1995;15:3672–3684.
25. Bito H, Deisseroth K, Tsien RW. CREB phosphorylation and dephosphorylation: a Ca^{2+}-and stimulus duration-dependent switch for hippocampal gene expression. Cell 1996;87:1203–1214.
26. Sakagami H, Kondo H. Immunohistochemical localization of Ca^{2+}/calmodulin-dependent protein kinase IV in mature and developing rat retina. Brain Res 1996;719:154–160.
27. Ochiishi T, Terashima T, Sugiura H, Yamauchi T. Immunohistochemical localization of Ca^{2+}/calmodulin-dependent protein kinase II in the rat retina. Brain Res 1994;634:257–265.
28. Yoshida K, Imaki J, Hujisawa H, Harada T, Ohki K, Matsuda H, Hagiwara M. Differential distribution of CaM kinases and induction of c-fos expression by flashing and sustained light in tat retinal cells. Invest Ophthalmol Vis Sci 1996;37:174–179.
29. Okuno S, Fujisasa H. Requirement of brain extract for the activity of brain calmodulin-dependent protein kinase IV expressed in *Escherichia coli.* J Biochem 1993;114:167–170.
30. Tokumistu H, Brickey DA, Glod J, Hidaka H, Sikela J, Soderling TR. Activation mechanisms for Ca^{2+}/calmodulin-dependent protein kinase IV: identification of a brain CaM kinase IV kinase. J Biol Chem 1994;269:28640–28647.
31. Tokumistu H, Enslen H, Soderling TR. Characterization of a Ca^{2+}/calmodulin-dependent protein kinase cascade: molecular cloning and expression of calcium/calmodulin-dependent protein kinase kinase. J Biol Chem 1995;270:19320–19324.
32. Okuno S, Kitani T, Fujisawa H. Evidence for the existence of Ca^{2+}/calmodulin-dependent protein kinase IV kinase isoforms in rat. J Biochem 1996;119:1176–1181.
33. Kitani T, Okuno S, Fujisawa H. Molecular cloning of Ca^{2+}/calmodulin-dependent protein kinase kinase β. J Biochem 1997;122:243–250.
34. Sakagami H, Saito S, Kitani T, Okuno S, Fujisawa H, Kondo H. Localization of the mRNAs for two isoforms of Ca^{2+}/calmodulin-dependent protein kinase kinases in the adult rat brain. Molec Brain Res (In press).
35. Sawamura Y, Sakagami H, Kondo H. Localization of mRNA for Ca^{2+}/calmodulin-dependent protein kinase I in the brain of developing and mature rats. Brain Res 1996;706:259–266.
36. Nakamura Y, Okuno S, Kitani T, Otake K, Sato F, Fujisawa H. Distribution of Ca^{2+}/calmodulin-dependent protein kinase kinase α in the rat central nervous system: an immunohistochemical study. Neurosci Lett 1996;204:61–64.

Dynamic Cells: Cell Biology of the 21st Century
S. Yagihashi, T. Kachi and M. Wakui, editors.

Shapes of osteoblasts and osteocytes tell their function and differentiation

Kazuhiro Abe, Hiroya Hashizume, Tatsuo Ushiki, Jun Kinoshita and Rieko Shibata
Department of Anatomy and Department of Orthopedic Surgery, Hokkaido University Graduate School of Medicine, Sapporo, Japan

Abstract. From scanning electron microscopic studies on mouse bones, we found that: 1) osteoblasts lining the bone matrix surface project large numbers of slender long cell processes, running parallel to each other, into the narrow spaces between the osteoblasts and the osteoid, 2) collagen fibrils on the osteoid just under the osteoblasts are formed running parallel to the osteoblast processes, 3) the osteoblast cell processes show many attachments with the tips of cell processes from osteocytes especially on the tibia, stress-loading bones, but the attachments are not frequent on the parietal bones, stress-unloading bones, and 4) osteocytes are already detectable within the osteoblast layer. The findings indicate that osteocytes are differentiated within the osteoblast layer and then buried into the bone matrix to form the meshwork to detect the mechanical stress loading to the bone. They communicate with the osteoblasts which determine the collagen fibril orientation of the bone lamellae for the stress by changing the orientation of the cell processes to decide the orientation of the collagen fibrils during the polymerization of collagen molecules.

Keywords: bone lamella, bone matrix, collagen fibrils, osteogenesis.

Introduction

Osteoblasts produce collagen to make bone matrix on bone, and it is said that osteoblasts become osteocytes after osteoblasts are embedded in the newly formed bone matrix. Osteoblasts have functional polarity: they make bone only on one side, the bony side. Is there any morphological specialization on the surface of osteoblasts facing the bone? Do osteoblasts transform or differentiate into osteocytes? These questions were answered according to our findings in scanning electron microscopic studies of bone cells.

The bone matrix mainly consist of densely packed collagen fibrils and minerals [1–5]. The arrangement of the collagen fibrils in the bone forms the bone lamellae of Haversian systems and the outer and inner circumferential lamellae. The lamellae are organized in relation to the mechanical strength loading to the bone [1,2,4,6–9]. Neighboring lamellae show a different orientation of collagen fibrils running parallel [10]. The orientation of collagen fibrils in the bone lamellae are considered to be determined at the bone surfaces, such as the outer bone

Address for correspondence: Prof Kazuhiro Abe, Department of Anatomy, Hokkaido University Graduate School of Medicine, North-15, West-7, Kita-ku, Sapporo 060-8638 Japan. Tel.: +81-11-706-5033. Fax: +81-11-706-7151. E-mail: kazuabe@med.hokudai.ac.jp surface under the periosteum, the inner bone surface under the endosteum, the inner surfaces of Haversian canals, and surfaces of bone trabeculae, when fibrilogenesis occurs under the osteoblasts.

Osteoblasts line the bone surface forming a cellular layer on the osteoid, the unmineralized bone matrix. Precise observations of the relationships between the shapes of the undersurfaces of osteoblasts and the surface of the osteoid just under the osteoblasts may explain functional characteristics of osteoblasts. Therefore, we tried to observe the arrangement of osteoblasts on the osteoid or bone, orientation of collagen fibrils seen on the bone surface, and the undersurface of the osteoblast layer on the osteoid, using scanning electron microscopes (SEM), in order to understand the functional role of osteoblasts in organizing the arrangements of collagen fibrils in bone on the surface of the osteoid. Osteocytes were also observed.

We used mouse bones, which are small enough and suitable for SEM, to observe wide surface areas for a better understanding of fundamental rules of cellular activities of bones. For SEM, the surface of the osteoid is exposed by pealing off the periosteum after fixing the tissue by glutaldehyde. The surface of the mineralized bone matrix is easily exposed by digesting soft tissues by immersing the bone in an NaHCl solution [11—13]. The upper surface of the osteoblasts can be exposed by careful peeling of the periosteum, keeping the osteoblasts under the periosteum after a glutaldehyde fixation. However, to expose the lower surface of osteoblasts and the osteocytes buried in the bone matrix, we have invented a new method to remove the bone matrix, consisting of collagen fibrils and hyrodoxy apatite, by decalcification with EDTA and digestion of collagen fibrils with a warm KOH solution, leaving the cellular elements [14].

Osteoblasts have a great number of long processes running parallel to orientate collagen fibrils running parallel in the bone lamellae

We mainly observed parietal bones with no loading of body weight and stress-loading bones in the extremities in young mice.

The bones, after being divested of covering soft tissues and cells, demonstrate the mineralized surface of the bone. The outer surface of the bones is smooth and illustrates the orientation patterns of superficial collagen fibrils in the mineralized bone matrix. The collagen fibrils run straight or curved and parallel to each other in a certain area, differently from neighboring areas (Fig. 1A). Round and oval osteocyte lacunae are scattered on the bone surface. Below the lacunae is usually flat and shows 10—30 pits for projection of cell processes from the osteocyte. Pits like those in the lacunae are scattered on the smooth bone surfaces. These pits are openings of canaliculi for the processes of osteocytes buried in the bone matrix.

The outer surface of the bones is lined by round, oval or polygonal osteoblasts of 10—25 μm in diameter, forming a cellular layer (Fig. 1B). Thus, the bone surfaces are completely covered by osteoblasts like epithelial cells. The osteoblast

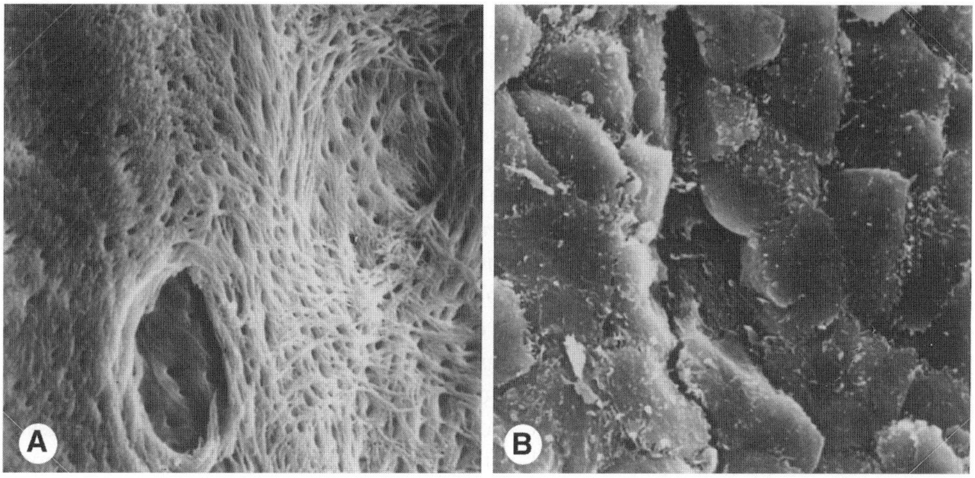

Fig. 1. Bone surfaces. **A:** Surface of the mouse tibia whose soft tissues have been removed by chemical digestion. Collagen fibrils in the matrix run parallel and mainly orient to the axis of the bone (up and down direction). **B:** Top view of the osteoblast layer on the tibia. Long axis of osteoblasts mainly directs to the axis of the bone (up and down direction).

surfaces directed to the periosteum are a little rounded and smooth. Small numbers of short microvilli appear on the smooth surface. The microvilli are only slightly frequent along the border with the neighboring osteoblasts.

The lower surface of the osteoblast layer facing the bone surface is well-demonstrated by exposure using our digestion techniques for the bone matrix (Fig. 2A).

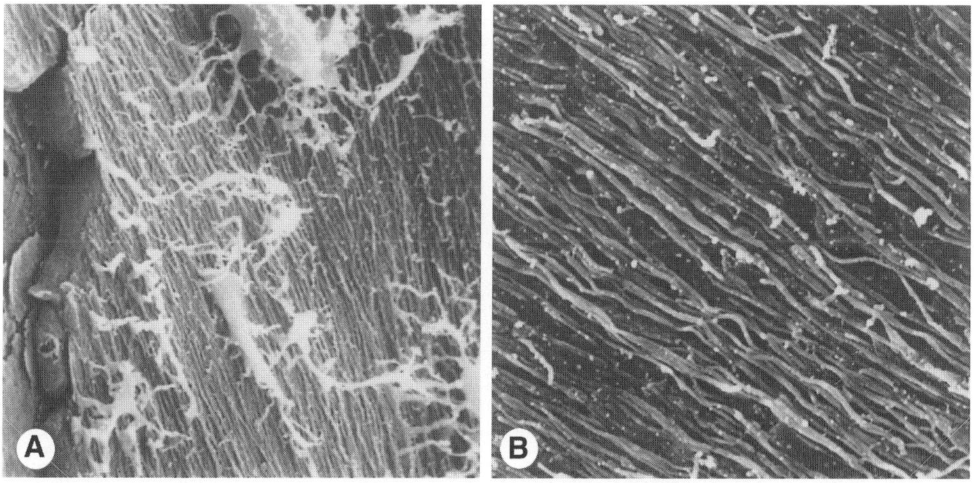

Fig. 2. Undersurfaces of the osteoblast layer. **A:** Peeled osteoblast layer. The lateral view of osteoblasts, lower surface of the osteoblast layer, and osteocytes are visible. **B:** Magnified images of the lower surface of the osteoblast layer. This is covered with many long osteoblastic processes running parallel.

The lower surface is flat and shows special differentiation of osteoblasts; it is covered with numerous long, thin, microvillous cell processes projecting from the osteoblasts. These processes are about 0.1 μm in thickness and straight or wavy (Fig. 2B). The processes run parallel in the same orientation in a certain area, but differently from the neighboring areas. Thus the lower surfaces of the osteoblast layers appear to be covered with a sheet of noodles running parallel. Such specialization of osteoblasts has never been noticed before.

The specimens observed in the top view of osteoblasts also illustrate the surfaces of the osteoid just under the osteoblasts at the areas where the osteoblasts have fallen off when the periosteum was dissected off. The surfaces of the osteoid reveal dense thin collagen fibrils running parallel and osteoblast processes which remained after removing the cell bodies of the osteoblasts (Fig. 3A). The cell processes run along the collagen fibrils. Occasionally, parallel osteoblast processes orient in a different direction from the fibrils around the processes, but one may find that the fibrils under or between the processes run parallel to the processes. Thus the osteoblasts extend long processes parallel to the collagen fibrils in a parallel orientation on the surface of the osteoid.

Osteoblasts regulate the orientation of collagen fibrils of the bone lamellae

In the tendon and cornea, composed of regularly and densely arranged collagen fibrils as well as bone, the regulation of the collagen fibril, bundle, and lamella formation has been explained in relation to the specialized structures of fibroblasts [18—22]. The fibroblasts during tendon morphogenesis have many narrow, deep cytoplasmic recesses which run parallel, contain single or small numbers of collagen fibrils, and fuse with the cell surface to form cellular folds and pro-

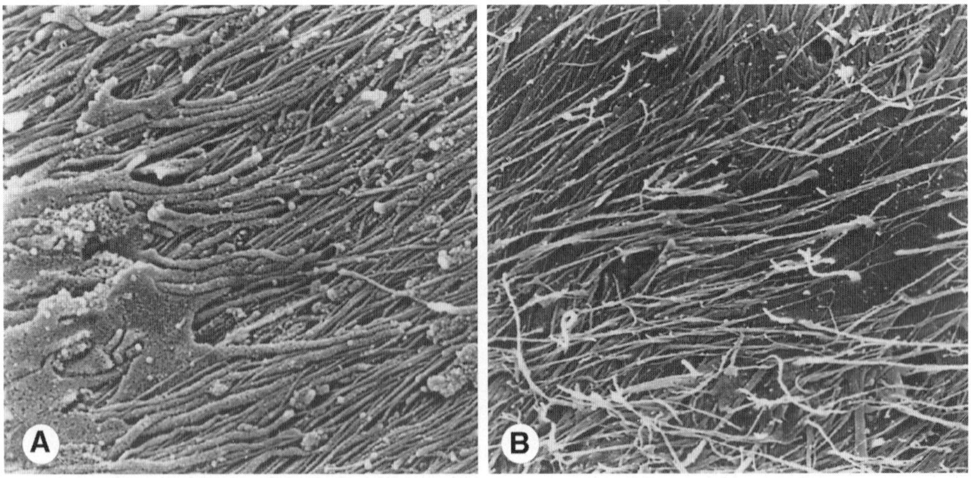

Fig. 3. Osteoid and osteoblast processes. **A:** Osteoblast processes run parallel to the collagen fibrils running parallel. **B:** Osteoblasts projecting parallel processes in two directions in two layers.

cesses enveloping forming fibril bundles; the cellular folds and processes also run parallel; thus, the bundles of collagen fibrils appear organized in parallel orientation in the tendon [15–18]. The fibroblasts in the embryonic cornea also have similar structures to those of tendon fibroblasts for fibrilogenesis [15]. The finding of tendon and cornea fibroblasts indicates that the collagen fibril orientation is regulated by long narrow cellular compartments assembling the fibrils. The tendon and cornea fibroblasts produce the fibrils around the cell bodies within the tendon and cornea.

However, differently from tendon and cornea fibroblasts, the orientation of collagen fibrils is always determined on the surface of the osteoid in collagen fibrilogenesis from procollagen molecules secreted by osteoblasts. The mechanism of the fibrilogenesis on the bone may be as described below.

Osteoblasts form a layer in close contact with the free surface of the bone or osteoid. The space between the lower surfaces of osteoblasts and the surface of the bone or osteoid is very narrow. Into this space, osteoblasts extend parallel with the long microvillous processes. The space between the lower surface of the osteoblasts and the surface of the osteoid is regarded as the collagen fibril forming compartment for the bone matrix. With reference to the concepts about the tendon and cornea, the long thread-like spaces or gaps between the osteoblast processes may provide the spaces where the procollagen molecules secreted by the osteoblasts are assembled to form a parallel arrangement of collagen fibrils. Then, the collagen fibrils may be piled up to form the osteoid, which is calcified from the bottom to transform into firm bone matrix.

Osteoblasts change the direction of cellular processes to form another lamella

Orientations of collagen fibrils are different between neighboring lamellae [10]. We have observed an osteoblast which projects two layers of cellular processes: a deeper layer of processes running parallel from an osteoblast is covered with another layer of processes running parallel in a different direction projecting from the same cell (Fig. 3B). Similarly, at the portion where a collagen fibril sheet runs in one direction on the deeper collagen fiber sheet running in another direction, the osteoblast processes run following the orientation of the covering fibrils (Fig. 3A). The findings indicate that another layer of bone lamellae is formed on the prior layer by changing the orientation of the osteoblast processes.

Such osteoblast processes are generally developed on young bones compared with bones from aged animals. The lamella formation may be more active when the bone is growing.

The bone lamellae are of plywood-like structure, which may strengthen the bone against mechanical stress [1,2,5–9]. The morphogenesis of the bone lamellae is considered to be related to the direction of the mechanical stress to the bone as shown by the longitudinal columnar arrangement of the osteons in the long bone. It is possible that the osteoblasts detect the direction of the stress to the bone and extend their long processes in relation to the direction of the stress.

However, the mechanical stress loaded to the parietal bone may not be great compared with the bones in the lower extremities as the surface of the parietal bone consistsof mosaic areas showing different orientations of parallel collagen fibrils.

Osteoblasts may receive information on mechanical stress to the bone from osteocytes

Stress-loading bones are focused on here in order to understand the mechanism whereby osteoblasts receive the information of strength loading to the bone.

The collagen fibrils in the bone matrix orient mainly in a longitudinal direction to the axis of the long bone. The osteoblasts covering the bone surface appear round or oval in the outer aspects. The long axis of the oval osteoblasts lies in various directions, but mainly longitudinally to the bone axis. Thus, the arrangement of osteoblasts appears to be related to the organization of stress-loading bones [19].

The lower surfaces of the osteoblastic layer on the tibia are also covered with a large number of long slender cell processes, oriented parallel and projecting from the osteoblasts. Differently from the parietal bone, however, the processes show many attachments with the terminals of cell processes from osteocytes under the osteoblast layer (Fig. 4A). On the bottom aspects of the peeled osteoblast layer, we can observe numerous torn-off remnants of the terminals on the osteoblast processes. The findings indicate that the cytoplasmic processes of osteocytes near the bone surface end to the side of the osteoblast processes.

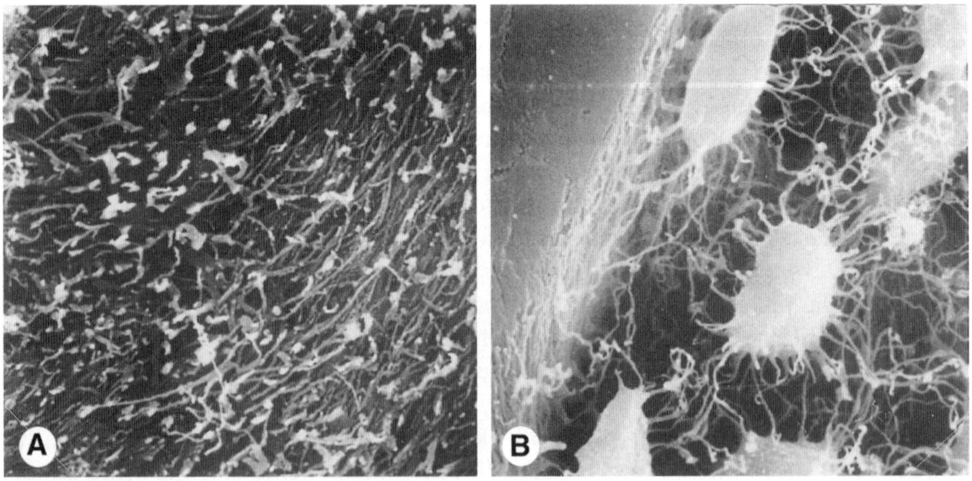

Fig. 4. Osteoblasts and osteocytes in the tibia. **A:** Lower surface of the osteoblast layer. There are many white dots which are the terminals of the osteocyte processes torn off when osteocytes have been removed. **B:** Lateral view of the osteoblast layer and osteocytes. The processes of osteocytes terminate at the lower surface of the osteoblast layer.

They are not frequent on the osteoblast processes on the parietal bone.

Osteocytes are well demonstrated in parietal bones and tibia by our digestion techniques. They project many long processes irregularly in three-dimensional directions in the bone. These cell processes are about 0.1 μm in diameter, of considerable length, sometimes branch, and terminate in the processes of neighboring osteocytes in an end-to-end, end-to-side or side-to-side juxtaposition, forming dense networks of cell processes in the bone. The osteocyte processes also reach to the free surfaces of the bone (Fig. 4B).

As described above, stress-loading bones and stress-unloading bones show a clear difference in the connection with osteocytes and osteoblasts by their cell processes. The stress-loading bones have a close connection between two types of cells. Such organization of these cells in bones suggests the possibility that the meshwork of osteocyte cell processes in the bone detects mechanical stress loading to the bone and works in communication with the osteoblast cell processes, which determine the collagen fibril orientation of the bone lamellae. It is assumed that the gap junctions between osteocyte processes are for routes of nutrients from the free surface to osteocytes buried in the deep bone matrix, but they also seem to play an important role as communicating junctions between osteocytes [1–3]. Our findings indicate that the junctions between osteocyte processes and osteoblast processes are firm; we have confirmed that at least some of these junctions are gap junctions.

It is possible that osteocytes receive information on the strength loading to the bone from their networks in the bone and transfer the information to the neighboring osteoblasts through the cell processes and the osteoblasts produce procollagen to form collagen fibrils for bone lamellae of architecture suitabe for the mechanical stress. Bone formation localized in a certain area may also be stimulated by stress loaded to the local osteocyte network system. In addition, osteocytes may provide information on the proper thickness of the lamellae, as suggested by the regular arrangements of osteocytes in the bone matrix.

Osteocytes differentiate in the osteoblast layer and are buried into the bone matrix

Observations of the lower surface of the osteoblast layer indicate differentiation of osteocytes.

A series of osteocytes burying into the bone matrix can be seen on the lower surface of the osteoblast layer (Fig. 5). Earlier osteocytes appear in the layer of osteoblasts; they are recognized as cells with cell processes projecting into the bone matrix, differently from the surrounding osteoblasts which extend the cell processes along the bone surfaces. Then they begin to be embedded into the matrix. Such cells are slightly protruded from the osteoblast layer on the bony side. Thus, it is clear that osteocytes are differentiated within the osteoblast layer and then buried into the bone matrix.

It has been considered that some osteoblasts gradually become enveloped by

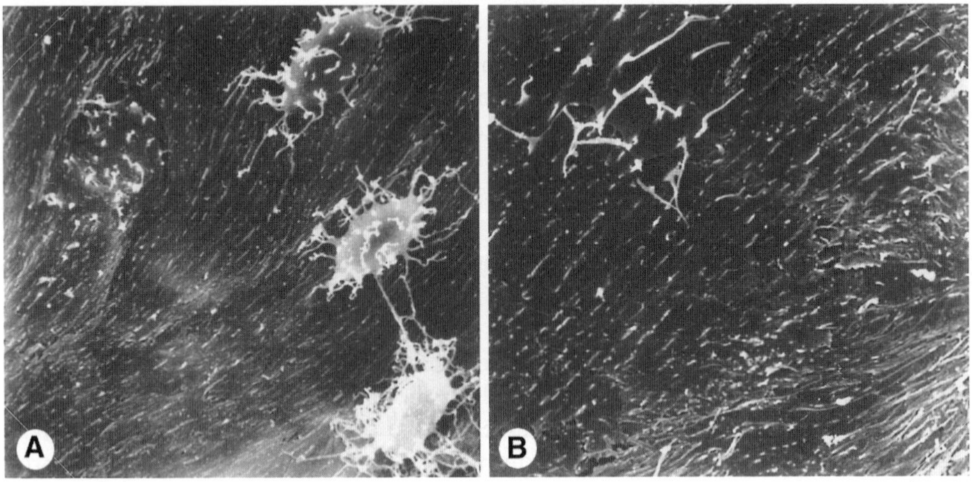

Fig. 5. Differentiation of osteocytes. **A:** Osteocytes which are differentiated in the layer of osteoblasts and then buried into the bone matrix. **B:** An osteocyte in the osteoblast layer.

their own secretions and are transformed into osteocytes, being buried in the newly formed bone matrix [1,3—5]. In this theory, the half-buried cells leaving the osteoblast layer are regarded as osteoblasts [20—23]. Osteoblasts densely cover the bone surface, but it cannot explain why only small numbers of osteoblasts become osteocytes. Our findings may answer this question as described below.

Osteocytes are differentiated from osteogenic progenitor cells in the layer of osteoblasts. The lower surfaces of the young osteocytes are attached by considerable numbers of cytoplasmic processes from osteocytes in the bone matrix. The points of the attachments are pulled as the surface bone lamellae is being thickened to move the osteocytes deeper, resulting in production and elongation of the cytoplasmic process from the new osteocytes. The new osteocytes begin to be buried into the osteoid and mineralized matrix after the new cell processes have reached a certain length. Thus, the osteocyte networks may be built up. Osteoblasts and osteocytic cells might be functionally different cells differentiated from the same osteogenic cells.

Conclusions

We first demonstrated by SEM that osteoblasts project numerous long processes running parallel to each other in the narrow space under the osteoblast layer and showed that shapes of osteoblasts and osteocytes tell their function and differentiation. The function and differentiation of osteoblasts and osteocytes that we considered are described below (Figs. 6 and 7).

1. Osteocytes are differentiated within the osteoblast layer covering the bone or osteoid and buried into the bone matrix.

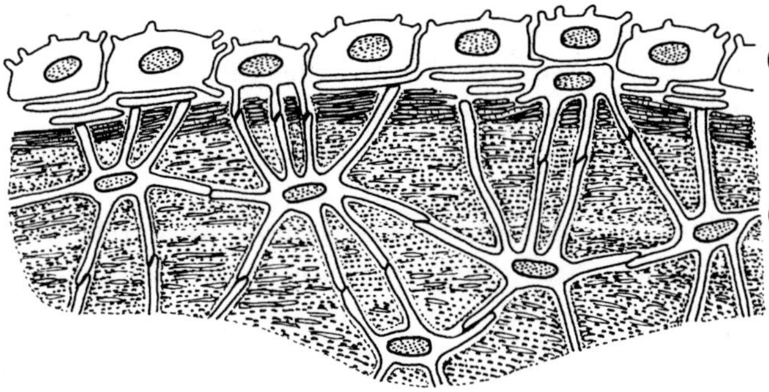

Fig. 6. Osteoblasts (Ob) and osteocytes (Oc). Osteocytes are differentiated within the osteoblast layer and then bury into the bone matrix. Osteocytes form a network system to perceive the stress-loading to the bone and inform the osteoblast of this through their cell processes. See the text for details.

2. Osteocytes form networks as a sensory system to perceive stress loading to the bone and also form a conducting system of the stress.
3. Osteocytes conduct the information of the stress loading to the bone through their cell processes to the local osteoblasts forming a cellular layer on the free surface of the bone.
4. Osteoblasts produce and secrete procollagen into the space between osteoblasts and the bone surface.
5. Osteoblasts also receive the information of collagen fibril arrangements in the bone lamella for making the fibrils in a reasonable direction for the stress and place their cell processes running parallel to the direction.
6. Narrow gaps or thread-like spaces are formed between the osteoblast processes in the narrow spaces under the osteoblasts and the procollagen mole-

Fig. 7. Osteoblasts. Osteoblasts receive information of the stress loading of the bone and thickness of the forming bone lamella from osteoblasts and extend the parallel cell processes to a certain direction to provide thread-like compartments for collagen fibril formation on the surface of the osteoid. Procollagen molecules are polymerized to form collagen fibrils along the osteoblast processes. Thus collagen fibrils running parallel in the bone lamellae are formed. See the text for details.

cules secreted from the osteoblasts are polymerized in the gaps to make collagen fibers running along the processes, resulting in the formation of collagen fibrils running parallel in the bone lamellae.

7. Osteoblasts know the proper thickness of the bone lamella and change the direction of the cell processes running parallel in the next lamella.

The above-mentioned conclusions may include many assumptions, but they appear reasonable from our findings. New approaches for understanding the functions of bone cells are encouraged for further studies.

References

1. Bloom W, Fawcett DW. A Textbook of Histology, 10th edn. Philadelphia: Saunders, 1975; 244–287.
2. Cormack DH. Ham's Histology, 9th edn. Philadelphia: Lippincott 1987;273–323.
3. Marks SC, Popoff SN. Bone cell biology; the regulation of development, structure, and function in the skeleton. Am J Anat 1988;183:1–44.
4. Matthews JL. Bone structure and ultrastructure. In: Urist MR (ed) Fundamental and Clinical Bone Physiology. Philadelphia: Loppincott, 1980;4–44.
5. Vaugman JM. The Physiology of Bone. Oxford: Clarendon Press, 1970;1–60.
6. Ascenzi A, Bonucci E. The tensile properties of single osteons. Anat Rec 1967;158:375–386.
7. Ascenzi A, Bonucci E. The compressive properties of single osteons. Anat Rec 1968;161: 377–392.
8. Boyde A, Riggs M. The quantitative study of the orientation of collagen in compact bone slice. Bone 1990;11:35–39.
9. Martin RB. Determinants of the mechanical properties of bones. J Biomech 1991;24(Suppl 1): 79–88.
10. Boyde A, Hobdell MH. Scanning electron microscopy of lamella bone. Z Zellforsch 1969;93: 213–231.
11. Abe K, Kanno T, Kitao K, Schneider GB. Sex differences in bone resorption: a scanning electron microscopic study of mouse parietal bones. Arch Histol Jpn 1984;47:429–440.
12. Abe K, Aoki Y. Sex differences in bone resorption in the mouse femur. A light- and electron-microscopic study. Cell Tiss Res 1989;52:299–310.
13. Abe K, Ohno K, Hasegawa K. Morphological relationships between osteoclasts and bone resorption surfaces on mouse parietal bones. Arch Histol Cytol 1990;53:179–186.
14. Abe K, Hashizume H, Ushiki T. An EDTA-KOH method to expose bone cells for scanning electron microscopy. J Electron Microsc 1992;41:113–115.
15. Birk DE, Trelstad RL. Extracellular compartments in matrix morphogenesis: collagen fibril, bundle, and lamellar formation by corneal fibroblasts. J Cell Biol 1984;99:2024–2033.
16. Birk DE, Trelstad RL. Extracellular compartments in tendon morphogenesis: collagen fibril, bundle, and macroaggregate formation. J Cell Biol 1986;103:231–240.
17. Squier CA, Bausch WH. Three-dimensional organization of fibroblasts and collagen fibrils in rat tail tendon. Cell Tis Res 1984;238:319–327.
18. Yang GH, Birk DE. Topographies of extracytoplasmic compartments in developing chick tendon fibroblasts. J Ultrastruct Molec Struct Res 1988;97:238–248.
19. Jones SJ, Boyde A, Pawley JB. Osteoblasts and collagen orientation. Cell Tiss Res 1975;159: 73–80.
20. Menton DN, Simmons DJ, Chang SL, Orr BY. From bone cell to osteocyte – an SEM study. Anat Rec 1984;209:29–39.
21. Nefussi JR, Sautier JM, Nicolas V, Forest N. How osteoblasts become osteocytes: a decreasing matrix forming process. J Biol Buccale 1991;19:75–82.

22. Palumbo C. A three-dimensional ultrastructural study of osteoid-osteocytes in the tibia of chick embryos. Cell Tiss Res 1986;246:125–131.
23. Palumbo C, Palazzini S, Zaffe D, Marotti G. Osteocyte differentiation in the tibia of newborn rabbit: an ultrastructural study of the formation of cytoplasmic processes. Acta Anat 1990; 137:350–358.

Dynamic Cells: Cell Biology of the 21st Century
S. Yagihashi, T. Kachi and M. Wakui, editors.

Intracellular calcium dynamics of secretory cells of the intestine

Yoh-ichi Satoh[1], Ken-ichi Kumagai[1], Hitoshi Satou[1] and Masayuki Orii[2]
Departments of [1]Functional Histology, and [2]Internal Medicine I (Hepato-gastroenterology), Iwate Medical University, Iwate, Japan

Abstract. *Background.* Digital imaging analysis for intracellular calcium ion, ($[Ca^{2+}]_i$), dynamics showed the heterogeneity of the responses of intestinal crypt cells to certain agonists. This study demonstrated the characteristics of $[Ca^{2+}]_i$ dynamics of some exocrine cells in the intestine.

Methods. Spatiotemporal dynamics of $[Ca^{2+}]_i$ were measured by digital imaging analysis in Fura-2/AM-loaded intact glandular end-pieces of pancreas, intestinal crypts, and Brunner's gland.

Results. Increased and rapid oscillatory fluctuations of $[Ca^{2+}]_i$ in mouse pancreatic acinus were elicited by agonists. In the crypts of the small intestine, carbachol (CCh)-induced $[Ca^{2+}]_i$ dynamics were restricted in Paneth cells, but goblet-cell reactions were not seen. Colon goblet cells in crypts reacted to cholinergic stimulus and to ATP, but no oscillatory fluctuation was observed. Brunner's gland of duodenum exhibited spontaneous oscillation of $[Ca^{2+}]_i$, even when no stimulant was applied. The persistent increase of $[Ca^{2+}]_i$ of the glands was caused by CCh.

Conclusions. Various exocrine cells in intestine possess their own stimulus-secretion coupling mechanism depending on $[Ca^{2+}]_i$ dynamics. Digital imaging analysis is useful to view the heterogeneity reflected with different physiological roles in intestinal tissues.

Keywords: digital imaging analysis, intestinal exocrine cells, intracellular calcium ion.

Introduction

Digital imaging analysis of intracellular calcium ion concentration ($[Ca^{2+}]_i$) dynamics is now a fundamental technique in cell biology. It is a useful way to distinguish the different responses to the transmitters/modulators in tissues consisting of various cells, such as intestinal epithelium [1]. Recently, $[Ca^{2+}]_i$ dynamics of epithelial cells in intact intestinal epithelium have been published [2,3]. Digital imaging analysis showed the heterogeneity of the responses of intestinal epithelium to different transmitters.

In this review, we focus our attention on the characteristics of $[Ca^{2+}]_i$ dynamics of some exocrine cells in the intestine: Paneth cells in the small intestine, goblet cells, and Brunner's gland in duodenum. Comparison with $[Ca^{2+}]_i$ dynamics of pancreatic acinar cells may refute the view that stimulus-secretion coupling mechanisms of different exocrine cells are similar, although $[Ca^{2+}]_i$ dynamics per se is an essential event in the secretion.

Address for correspondence: Y. Satoh MD, Department of Functional Histology, Iwate Medical University, Uchimaru 19-1, Morioka, Iwate 020-8505, Japan. Tel.: +81-196-51-5111 (ext. 3351); Fax: +81-196-51-5605. E-mail: yisatoh@iwate-med.ac.jp

Preparation of the specimens for digital imaging analysis

We offer some suggestions for preparation of intact intestinal epithelial specimens for digital imaging of $[Ca^{2+}]_i$ dynamics. First, to digest the connective tissue, purified collagenase is recommended because most crude collagenase possesses tryptic or protease activities. Second, the isolated tissue specimens must be immediately put on the cover glass in a perifusion chamber because epithelium in floating condition deteriorates rapidly. Third, we avoided the isolation procedure using Ca^{2+}-free medium [3]. It is doubtful that the cells once exposed to a Ca^{2+}-free environment retain normal $[Ca^{2+}]_i$ response.

For measurement of free ions, many fluorescent dyes have been introduced [4]; Fura-2 and Indo-1 are widely used because radiometry can determine semiquantitatively the $[Ca^{2+}]_i$ of living cells [5]. At digital imaging it is vital to be aware of the possibility of artifact at loading of the dye (e.g., compartmentalization) and to keep the optical system meticulously clean.

Pancreatic acinar cells

Because many studies are available [6], the characteristic features of $[Ca^{2+}]_i$ dynamics of pancreatic acinar cells are briefly summarized here. Physiological concentrations of cholecystokinin (CCK) and acetylcholine induce repetitive $[Ca^{2+}]_i$ spikes of each cell, commonly termed Ca^{2+} oscillation. Higher concentrations evoke biphasic $[Ca^{2+}]_i$ dynamics; with a rapid transient peak (steep phase) followed by a sustained increase (plateau phase). It is noticeable that agonist-induced $[Ca^{2+}]_i$ increase (so-called intracellular Ca^{2+} wave) initiate from apical secretory pole and propagate toward basal region in isolated acinar cells. However, the Ca^{2+} wave in acinar cells of intact acinus covered by basement membrane initiate from basal region and propagate toward apical pole [7].

Paneth cells

Paneth cells, located in the intestinal crypt bottom, are characterized by the eosinophilic secretory granules, containing antibacterial agents (lysozyme and defensin/corticostatin) [8]. Paneth cell secretion after changes in the intestinal bacterial milieu was inhibited by atropine, a muscarinic antagonist [9]. In accordance with this result, we demonstrated that carbachol (CCh) induced $[Ca^{2+}]_i$ dynamics in Paneth cells [2]: after CCh stimulation (3—10 μM), $[Ca^{2+}]_i$ increased abruptly (steep phase) and then was sustained even after cessation of CCh stimulation but declined gradually (plateau phase). The steep phase of $[Ca^{2+}]_i$ results mainly from Ca^{2+} release from the intracellular Ca^{2+} store and partially from Ca^{2+} influx from extracellular fluid; the plateau phase proceeds almost exclusively from a Ca^{2+} influx. CCh sometimes elicited slow (duration of a few minutes) sinusoidal fluctuations.

Goblet cells

Goblet cells are mucous-secreting exocrine cells in epithelium. Previous studies revealed that secretion of goblet cells was induced by cholinergic stimuli [10]. CCh induced the persistent $[Ca^{2+}]_i$ increase in colon crypts, which were composed principally of goblet cells. The increase was in whole crypts; however, crypt bottom showed more prominent and persistent increases (Fig. 1). Extracellular ATP caused monophasic transient $[Ca^{2+}]_i$ increase [3]. In both CCh- and ATP-induced $[Ca^{2+}]_i$ dynamics, no spike oscillation was detected. Nicotine, 5-HT, CCK, histamine, choleratoxin and guanylin have no effect on $[Ca^{2+}]_i$ dynamics of goblet cells of mouse and human. However, mucous contents of goblet cells vary in different intestinal regions and in different species. Similarly, the response of goblet cells to certain transmitters will be considerably different.

Brunner's gland

In the proximal small intestine of many mammals, seromucous-secreting cells cluster and form exocrine glands, which were first described by Brunner [11].

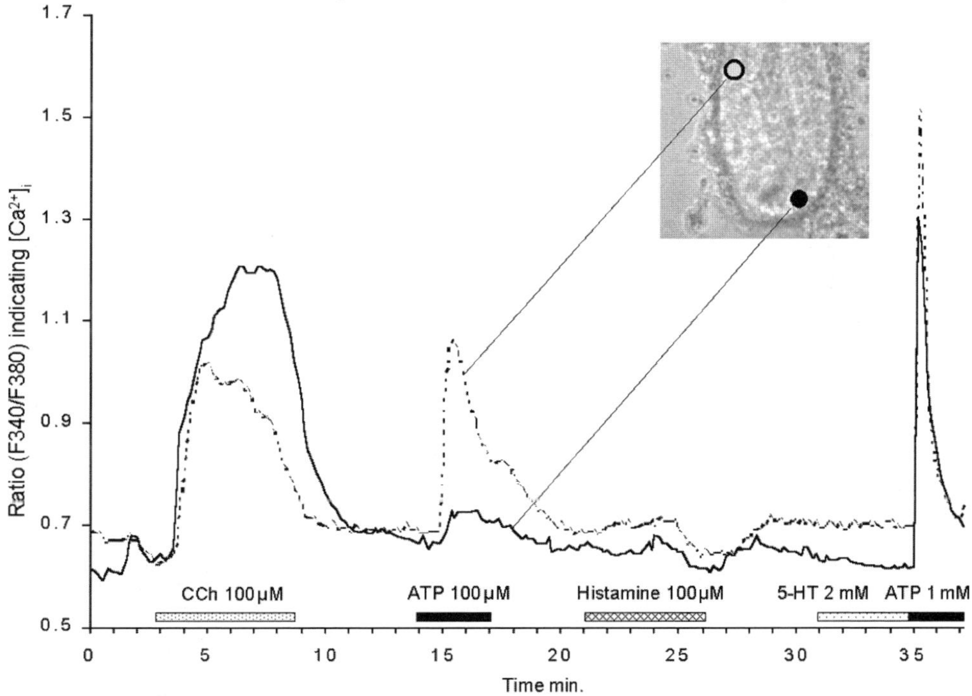

Fig. 1. $[Ca^{2+}]_i$ dynamics of an intact human colon crypt. Graph shows temporal changes of $[Ca^{2+}]_i$, measured by Argus 50/CA (Hamamatsu Photonics, Hamamatsu, Japan). Note that the carbachol (CCh)-induced $[Ca^{2+}]_i$ reaction is predominant in the lower region (solid line), whereas the ATP-induced reaction is prominent in the upper region (dotted line). Histamine and 5-HT have no effect. Measured area: 4 μm^2.

The most noticeable characteristic of $[Ca^{2+}]_i$ dynamics of the mouse Brunner's gland is spontaneous oscillatory fluctuation. Repetitive $[Ca^{2+}]_i$ spikes of each cell appeared without any secretagogues (Fig. 2A). The Ca^{2+} oscillation of each cell was not fully synchronized; however, octanol (1 mM; a putative blocking agent of gap junctions) decreased the frequency of the oscillation. Removing extracellular Ca^{2+} considerably diminished the oscillation. Spontaneous oscillations of $[Ca^{2+}]_i$ have been observed in many excitatory cells [12]; however, membrane depolarization by high K^+ (56 mM) had no effect on the $[Ca^{2+}]_i$ dynamics of Brunner's gland cells.

Spontaneous oscillation may indicate that Brunner's gland secretes actively even in a resting state without any stimulus from enteric nerves. This view can be supported by the small number of secretory granules stored in apical regions of mouse Brunner's gland. The glands secrete the basic fluid for neutralization of acidic chyme, which constantly enters from the stomach [13]. Therefore, a ceaseless secretory mechanism may be suitable for the gland's role in vivo. In contrast, pancreatic acinar cell secretion can be triggered almost solely by external neurohormonal stimuli because the digestive enzymes are needed not constantly but intermittently.

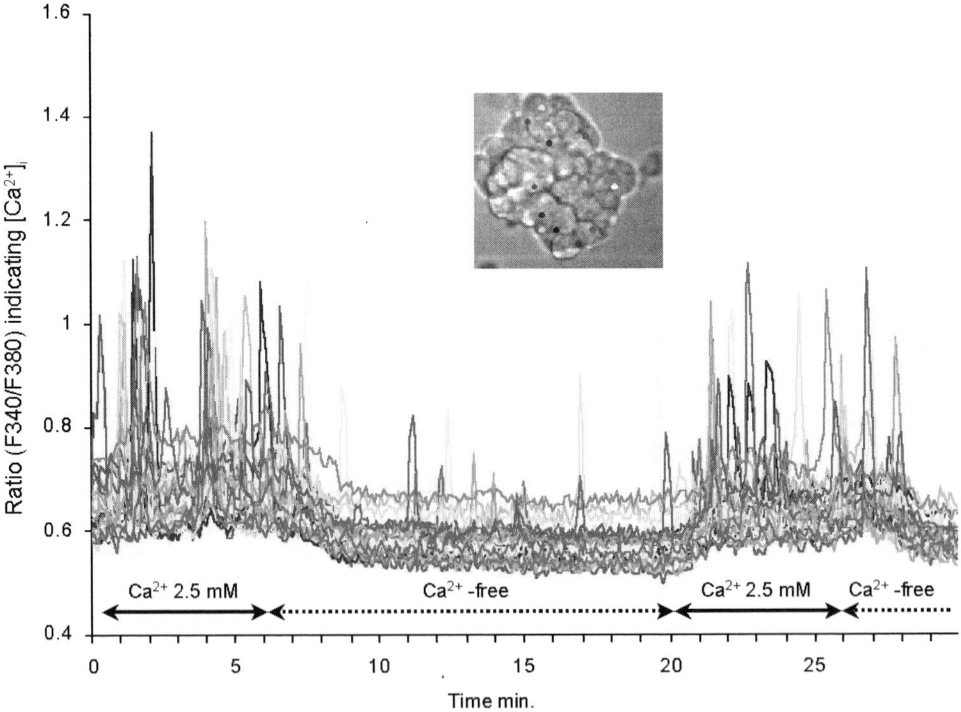

Fig. 2A. $[Ca^{2+}]_i$ dynamics of a mouse Brunner's gland: spontaneous oscillation of $[Ca^{2+}]_i$ with or without extracellular Ca^{2+}. Measured area: 4 μm^2.

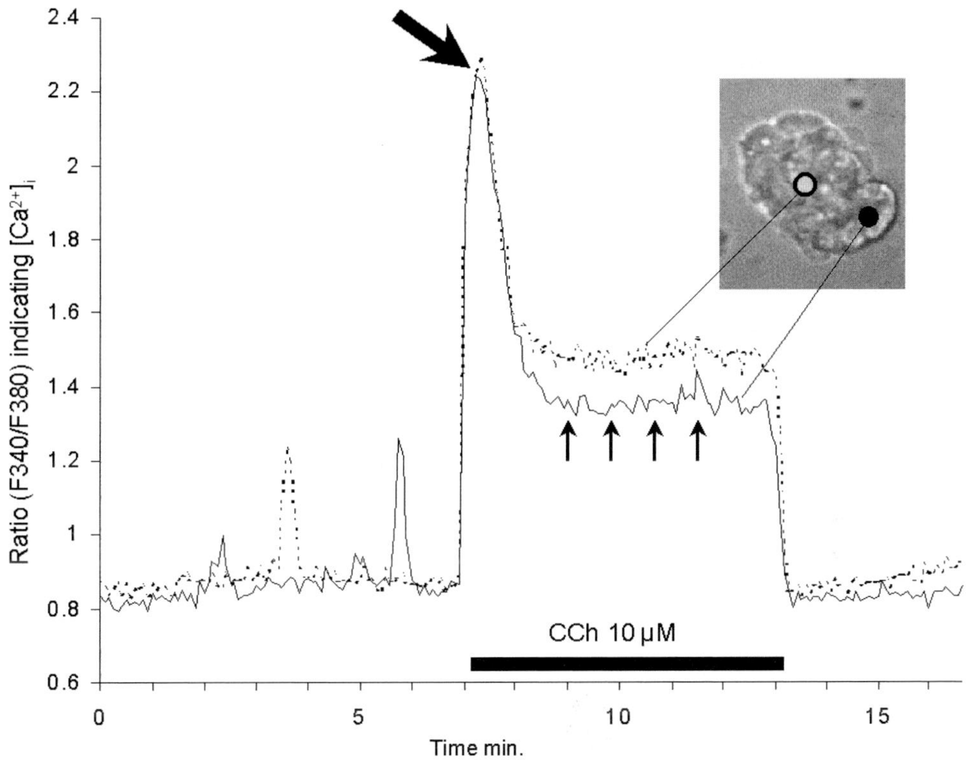

Fig. 2B. $[Ca^{2+}]_i$ dynamics of a mouse Brunner's gland: CCh-induced $[Ca^{2+}]_i$ dynamics, showing steep (thick arrow) and plateau phases (thin arrows). Measured area: 4 μm^2.

In the pancreatic acinar cells, as described above, the initial point of the Ca^{2+} wave is controversial [7]. Based on the observations on spontaneous oscillations of intact Brunner's gland acini, the Ca^{2+} wave was initiated from both regions; in many cases from basal to apical, and sometimes from apical to basal.

CCh (1–10 μM) elicited $[Ca^{2+}]_i$ dynamics of entire glandular end-pieces; an initial steep phase and sustained plateau phase were observed (Fig. 2B). Lower CCh concentrations (0.1 μM) increased the frequency of oscillations. Neither sodium fluoride, ATP (10 μM), nor atropine (100 μM, a muscarinic antagonist) affected the $[Ca^{2+}]_i$ dynamics.

Conclusion

Secretion is an essential function of living cells, and it is accompanied by $[Ca^{2+}]_i$ dynamics. However, the digital imaging analysis on $[Ca^{2+}]_i$ dynamics of intact tissue specimens revealed important differences and specialization among the exocrine cells in intestine (Table 1). This is only the beginning of studying the variety of $[Ca^{2+}]_i$ dynamics. To understand life, we first must understand the phy-

Table 1. Effects of CCh, CCK and ATP on $[Ca^{2+}]_i$ dynamics of exocrine cells.

	Without stimulants	CCh	CCK	ATP
Exocrine pancreas	No dynamics	Spike oscillation (low), biphasic increase (high)	Spike oscillation biphasic increase (high)	No dynamics
Paneth cells	No dynamics	Biphasic increase, sinusoidal oscillation	No dynamics	No dynamics
Goblet cells (colon[a])	No dynamics	Biphasic increase	NE	Monophasic transient increase
Brunner's gland of duodenum	Spontaneous spike oscillation	Increase frequency of oscillation (low), biphasic increase (high)	NE	No dynamics

[a]Human material (biopsy specimen at colonoscopy); others are from mice. ATP, adenosine triphosphate; CCh, carbamylcholine; CCK, cholecystokinin; NE, not examined.

siological responses in living tissues, which requires more than the study of only isolated and culture cells.

Acknowledgements

This work was supported by a research grant from the Ministry of Education, Culture and Science of Japan (08457007).

References

1. Furness JB, Costa M. The enteric nervous system. Edinburgh: Churchill Livingstone, 1987.
2. Satoh Y, Habara Y, Ono K, Kanno T. Carbamylcholine- and catecholamine-induced intracellular calcium dynamics of epithelial cells in mouse ileal crypts. Gastroenterology 1995;108:1345–1356.
3. Leipziger J, Kerstan DM, Nitschke R, Greger R. ATP increases $[Ca^{2+}]_i$ and ion secretion via basolateral P2Y-receptor in rat distal colonic mucosa. Pflügers Arch 1997;434:77–83.
4. Haugland RP. Handbook of fluorescent probes and research chemicals. 6th edn. Eugene: Molecular Probes, 1996.
5. Shotton DM. Electronic light microscopy: present capabilities and future prospects. Histochem Cell Biol 1995;104:97–137.
6. Habara Y, Kanno T. Stimulus-secretion coupling and Ca^{2+} dynamics in pancreatic acinar cells. Gen Pharmacol 1994;25:843–850.
7. Kanno T, Habara Y, Satoh Y. Ca^{2+} wave propagation in cells of pancreatic acinus reverses after cell dispersion. Biomed Res 1991;12:429–434.
8. Ouellette AJ, Greco RM, James M, Frederick D, Naftilan J, Fallon JT. Developmental regulation of cryptdin, a corticostatin/defensin precursor mRNA in mouse small intestinal crypt epithelium. J Cell Biol 1989;108:1687–1695.
9. Satoh Y. Atropine inhibits the degranulation of Paneth cells in ex-germ-free mice. Cell Tiss Res 1988;253:397–402.
10. Specian RD, Neutra MR. Regulation of intestinal goblet cell secretion. I. Role of parasympathetic stimulation. Am J Physiol 1982;242:G370–G379.

11. Brunner JC. De glandulis in intestio duodeno hominis detectis. University of Heidelberg, Heidelberg, 1688 (Dissertation).
12. D'Andrea P, Zacchetti D, Meldolesi J, Grohovaz F. Mechanism of $[Ca^{2+}]_i$ oscillations in rat chromaffin cells. J Biol Chem 1993;268:15213–15220.
13. Lang IM, Tansy MF. Brunner's glands. In: Young JA (eds) Gastrointestinal Physiology IV, International Review of Physiology. Baltimore: University Park Press, 1983;28:85–102.

1998 Elsevier Science B.V.
Dynamic Cells: Cell Biology of the 21st Century
S. Yagihashi, T. Kachi and M. Wakui, editors.

ATP-sensitive K^+ channels in β-cells; their role in stimulus-secretion coupling

Carina Ämmälä[1], Noritaka Kamimura[2] and Mark J. Dunne[2]
[1]Institute of Physiology and Pharmacology, Department of Medical Biophysics, Göteborg University, Göteborg, Sweden; and [2]The Department of Biomedical Science, The University of Sheffield, Sheffield, UK

Abstract. Here we describe the relationship between the control of K_{ATP} channel function in pancreatic β-cells and the regulation of insulin secretion. Focus is placed upon two areas of complementary activity in the field of ion channels and stimulus–response coupling in β-cell. First, the molecular biology of the β-cell K_{ATP} channels, where studies have demonstrated that the channel is a novel heteromultimeric complex of an ATP-binding cassette protein, designated SUR1, and a K^+ channel. And secondly, we review the evidence that pathophysiology of the β-cell occurs in association with loss of K_{ATP} channel function. In this part of the review we document how gene variations in either of the channel subunits leads to unregulated, and inappropriately high rates of insulin secretion in humans, and the condition persistent hyperinsulinaemic hypoglycaemia of infancy. It is now clear that not only do K_{ATP} channels govern the onset of depolarisation-response coupling in β-cells, but that they also play a role in insulin-secreting cell survival by preventing insulin hypersecretion and the onset of apoptosis.

Keywords: insulin secretion, K_{ATP} channel, sulphonylurea receptors.

Introduction

Understanding human physiology is invariably advanced by resolving the pathogenesis of inherited diseases. Over the past few years it is becoming increasingly obvious that a number of clinical disorders can be directly correlated with abnormalities of cell membrane ion channels. The best described example of aberrant changes in ionic conductance leading to a disease phenotype is cystic fibrosis in which over 500 different mutations in the gene encoding the cystic fibrosis transmembrane conductance regulator protein, a cAMP-gated chloride channel, have so far been described. Recently, the first ATP-sensitive potassium (K_{ATP}) channel disease has been described. This occurs in association with the potentially lethal infantile disorder persistent hyperinsulinaemic hypoglycaemia of infancy (PHHI), a condition resulting from the loss of the ionic control of pancreatic β-cell function leading to unregulated insulin secretion. Characterization of stimulus-secretion coupling events in PHHI tissue has provided a unique

Address for correspondence: Dr C. Ämmälä, Institute of Physiology and Pharmacology, Department of Medical Biophysics, Göteborg University, Medicinaregatan 11, P.O. Box 433, SE 405 30 Göteborg, Sweden. Tel.: +46-31-773-3559. Fax: +46-31-773-3558. E-mail: ammala@mednet.gu.se insight into the normal physiological control mechanisms that govern secretion from pancreatic β-cells.

Background

Pancreatic β-cells are electrically active, and in the presence of stimulatory concentrations of glucose (typically greater than 5.5 mM) they will depolarise, and generate cycles or oscillations of depolarisation/repolarisation upon which Ca^{2+}-dependent action potentials are superimposed. These action potentials facilitate accelerated voltage-gated Ca^{2+} influx which then triggers Ca^{2+}-regulated exocytosis whereby secretory granules fuse with the plasma membrane in order to release their contents.

The first observations that initiated this field were made almost 30 years ago when, using glass microelectrodes, Dean and Matthews [1] were able to show that, in the presence of carbohydrate secretagogues, β-cells generate spike potentials. The link between this finding and the concept that calcium ions are crucial for insulin release was subsequently made by Matthews and Sakamoto [2], who demonstrated that these voltage-gated spikes were caused by the inward movement of calcium ions. Later, it was also shown, using radioisotope studies, that carbohydrate metabolism is coupled to the initiation of the membrane depolarisation by a decrease in the permeability of the membrane to potassium ions. However, it was not until the mid-1980s and the application of the patch-clamp technique to isolated cells, that these events were further linked to the molecular level that we now understand to be important for stimulus–response coupling events. Such electrophysiological experiments have not only revealed the involvement of plasma membrane ion channels in the control of cell function, but a modification in the patch-clamp technique is now unravelling the molecular events that control the final steps in the release of insulin from the pancreatic β-cells.

Consensus model for the ionic control of insulin release

Based upon direct recordings of ion channel events from insulin-secreting cells, our current hypothesis to account for how glucose causes changes in a cell membrane potential has two major features. First, that the resting membrane potential is maintained by the electrogenic Na^{+}-K^{+} ATPase and open K_{ATP} channels [3], which allows K^{+} to exit the β-cells down the electrochemical gradient. Secondly, the channels act as "metabolic sensors" and close as a consequence of the increase in intracellular ATP/ADP ratio produced by enhanced glucose metabolism in response to an elevated blood sugar level after a meal [4]. K_{ATP} channel inhibition causes a depolarization of the cell membrane leading to the activation of voltage-dependent Ca^{2+} channels and the ensuing rise in the free intracellular concentration of Ca^{2+} ($[Ca^{2+}]_i$) close to the membrane then initiates the release of insulin. In this scheme (Fig. 1) voltage-dependent Ca^{2+} influx is shown as

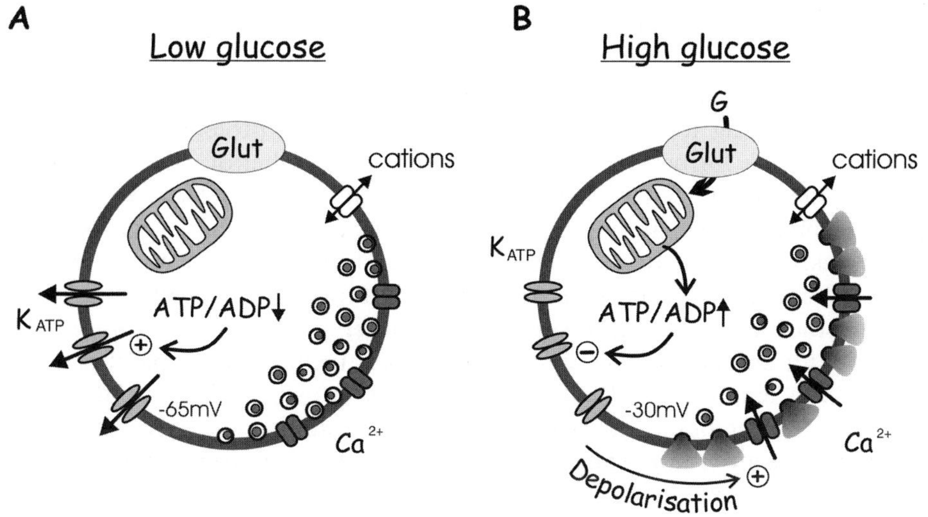

Fig. 1. Stimulus-secretion coupling in pancreatic β-cells. **A:** At nonstimulatory glucose concentrations the β-cell resting membrane potential is determined by open K_{ATP} channels. **B:** When the glucose concentration is elevated the stimulated metabolism increases the ATP/ADP ratio and K_{ATP} channels close. This leads to membrane depolarisation and the opening of voltage-activated Ca^{2+} channels. The increased Ca^{2+} influx then initiates the release of insulin through exocytosis of the secretory granules.

the dominant influence upon the release of insulin-containing granules by exocytosis. A rise in $[Ca^{2+}]_i$ is therefore critically dependent upon the activity of K_{ATP} channels and the pivotal role that these have in β-cell stimulus–response coupling mechanisms. The model also iterates that the pharmacological regulation of K_{ATP} channels has important medicinal implications. So, for example, compounds that will selectively inhibit these channels such as the antidiabetic sulphonylureas (glibenclamide, tolbutamide, etc.) will mimic the actions of glucose and promote insulin release, whilst agents that act as K_{ATP} channel agonists (openers), such as diazoxide, exert the opposite effect and tend towards the inhibition of secretion by reversing the glucose-dependent depolarisation of the membrane potential.

Despite the fact that this model is now generally accepted as an integral component of early signalling events in β-cells, it is not until now that there has been evidence to show that when these events are altered the control of insulin secretion from β-cells is lost. In this respect, two major advances of significant importance to this field have been made over the past two years, findings that now provide convincing support for the consensus model of β-cell stimulus–response coupling (Fig. 1). First, the molecular architecture of the β-cell K_{ATP} channel has been partially resolved [5,6], and second, studies of the ionic control of secretion in PHHI, a rare neonatal disorder, have revealed that uncontrolled hypersecretion of insulin occurs in association with the loss of functional K_{ATP} channels in β-cells [7–10].

Architecture of the β-cell K_{ATP} channel

Insulin-secreting cells express a K_{ATP} channel complex formed by the interaction of subunits belonging to at least two distinct families of proteins (Fig. 2). The K^+-selective pore is formed by the Kir6.2 subunit, a member of the inward rectifier K^+ channel family. Kir6.2 has a predicted membrane topology with two α-helical transmembrane-spanning domains linked by a highly conserved sequence of amino acids. As this domain shares sequence homology with the P- or H5 region of K_V channels, it is thought to be important for the control of K^+ selectivity through the pore. The other subunit is a larger protein – a receptor with high affinity for sulphonylureas, designated SUR1. In man this subunit is 1,581 amino acids in length and like the cystic fibrosis transmembrane conductance regulator, it is a member of the superfamily of ATP-binding cassette (ABC) proteins. These proteins share common structural homology with numerous predicted transmembrane sequences and nucleotide-binding domains (NBDs). Neither the native Kir6.2 subunit nor the SUR1 subunit will form operational K^+ channels independently. However, when coexpressed, K^+ channel currents are generated that closely resemble those of the native K_{ATP} channel complex [5,6]. Current research, addressing the topological organisation of the K_{ATP} channel, suggests that it is an octameric complex with a 1:1 stoichiometry formed by four Kir6.2 subunits lining the pore coupled to four SUR1 subunits [11,12]. Ion selectivity, rectification and gating of the complex appear to be determined by Kir6.2 [13], whereas SUR1 serves as a regulatory unit of the K^+ channel by conferring high sensitivity on the key pharmacological regulators diazoxide [14] and the antidiabetic sulphonylureas [15]. Channel sensitivity to adenine and guanosine nucleotides is a complex process involving both subunits. Thus channel activation by MgADP, MgGDP and MgGTP is mainly conferred by SUR1 through interactions with the NBD regions [14,16,17], whereas ATP inter-

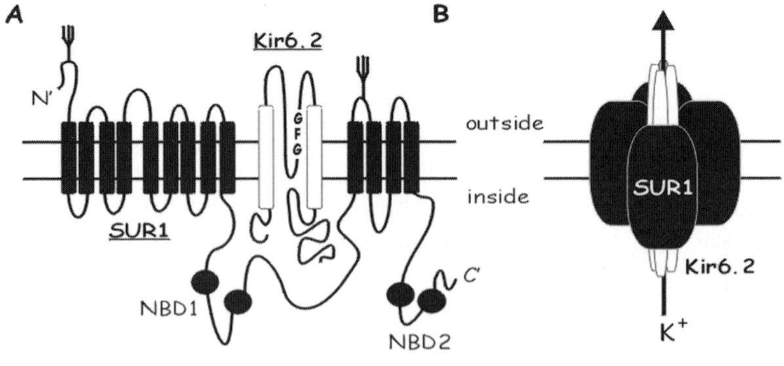

Fig. 2. Architecture of the β-cell K_{ATP} channel. **A:** The K_{ATP} channel is composed of at least two different subunits, Kir6.2 and SUR1. Kir6.2 belongs to a family of inwardly rectifying K^+ channel, whereas SUR1 is a member of the ABC transporter family. **B:** The channel is a multimeric protein complex consisting of four Kir6.2 subunits coupled to four SUR1 subunits.

acts with both SUR1 and Kir6.2 [18,19] to inhibit channel activity.

The gene that encodes SUR1 consists of 39 exon boundaries and is clustered with the Kir6.2 gene — a single open reading frame lying immediately 3′ of the SUR1 gene separated by only 4.5 kilo base pairs [5]. The location of these genes in man on the short arm of chromosome 11 corresponds to a genetic locus "linked" to familial forms of the childhood disorder PHHI [20,21]. We now have very good evidence that gene defects encoded within SUR1/Kir6.2 can be correlated with the functional loss of K_{ATP} channels in β-cells and that this leads to loss of regulated insulin secretion and the pathogenesis of this potentially lethal infantile disease [7—10].

Persistent hyperinsulinaemic hypoglycaemia of infancy (PHHI)

Hypoglycaemia is a relatively common childhood metabolic abnormality and when persistent or recurrent is most frequently a consequence of hyperinsulinism; PHHI. Until recently the pathophysiology of PHHI was not understood, although a defect in the function of pancreatic β-cells was first alluded to more than 15 years ago when it was shown that glucose failed to promote a concentration-dependent release of insulin from PHHI β-cells in vitro [22]. Sporadic cases of PHHI are thought to be rare in the general population (~1/45,000), but the disease has an incidence in communities with high rates of consanguinity that approaches that of cystic fibrosis in European Caucasians, i.e., approximately 1/2,500 live births. PHHI has a variable clinical phenotype with mild vs. severe, early onset vs. late onset, and familial vs. nonfamilial conditions. Severe forms are present within the first few hours/days following birth, as the result of extreme and sustained low blood glucose levels due to unregulated hypersecretion of insulin. Failure to recognise and to promptly treat hypoglycaemia carries a substantial risk of severe brain damage and mental retardation because of the lack of fuel that is needed to sustain normal brain metabolism. Medical therapy for the disorder involves an increased carbohydrate intake to meet the elevated requirement, and usually one or more drugs that inhibit insulin secretion. These agents include diazoxide which was first introduced to treat hyperinsulinism in the 1960s, and somatostatin. The mechanism of action of both of these agents has been shown to involve K_{ATP} channel activation in normal β-cells, which results in the inhibition of insulin secretion since increases in K^+ channel open events cause a hyperpolarisation of the membrane potential, closing the voltage-dependent Ca^{2+} channels. Unfortunately, however, the responsiveness of children with PHHI to these agents is highly variable, ranging from extreme sensitivity through to total drug resistance. Those patients who do not show adequate and immediate response to K^+ channel openers (alone or in combination with other compounds) require surgery to remove their pancreas — usually a 95% pancreatectomy, in order to prevent recurrent hypoglycaemia. The variability in sensitivity to medical therapy has until now been unexplained (see below). Pancreatectomy usually induces a clinical remission of symptoms, but this procedure will invariably predispose the PHHI patient to the development of diabetes in later life.

Impaired control of ion channels in β-cells and hypersecretion of insulin

The critical link between understanding how the structure/function of K_{ATP} channels is associated with the clinical manifestation of PHHI came as a result of recent experiments studying the in vitro function of acutely isolated β-cells from patients affected with this condition. In the experiments insulin-secreting cells were prepared from several children with sporadic and familial disease following pancreatectomy. Using patch-clamp techniques under a number of experimental conditions, it was revealed quite remarkably 1) that β-cells from the children were "spontaneously" electrically active even in the presence of low ambient glucose concentrations, and 2) that the appearance of action potentials was brought about in association with the absence of functional K_{ATP} channels [7—10]. Recently, it has also been shown that PHHI is caused by the *selective* loss of K_{ATP} channel function at the cellular level since the major regulatory and pharmacological properties of delayed-rectifier K^+ channels, Ca^{2+} and voltage-gated K^+ channels and nonselective cation channels are apparently unaffected by the disease phenotype in PHHI β-cells [23].

How the absence of K_{ATP} channels can result in such a major change in the function of β-cells is readily understood when one considers the fundamental role these proteins have in the control of β-cell membrane potential (Fig. 3). Thus, in isolated PHHI β-cells the resting membrane potential is close to the

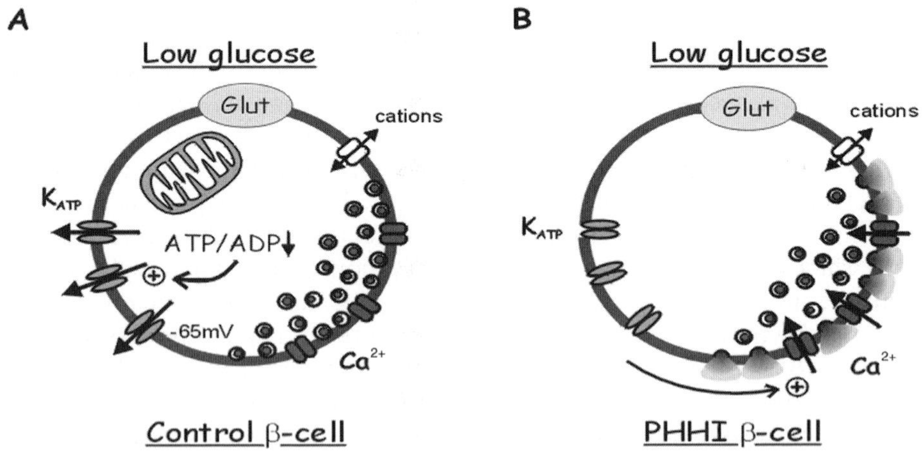

Fig. 3. Unregulated insulin secretion in PHHI. **A:** In normal β-cells open K_{ATP} channels are responsible for maintaining the membrane at hyperpolarized potentials in the resting state. **B:** As PHHI β-cells lack functional K_{ATP} channels, the membrane potential will remain depolarised even at non-stimulatory glucose concentrations, voltage-activated Ca^{2+} channels are open and as a consequence lead to the persistent release of insulin.

threshold for the activation of voltage-dependent Ca^{2+} channels, and it is activity from these channels that leads to the appearance of spontaneously regenerating action potentials in the "unstimulated" PHHI β-cell. Since Ca^{2+} influx is a key determinant of insulin secretion under normal conditions it has been proposed that inappropriate Ca^{2+} channel activity accounts for the unregulated secretion of insulin [8,9] (Fig. 3).

The correlation of K_{ATP} channel defects with PHHI

Several recent papers have described mutations in the SUR1 and Kir6.2 genes associated with PHHI. Two variations in the Kir6.2 gene have been reported [24,25] and 7 mutations in SUR1 have been described, three of which encode putative premature truncations of the protein [26—29]. However, with the exception of the exon 35 SUR1 gene mutation [9,26], none of the mutations have so far been found in patients where it has been possible to obtain isolated β-cells for functional studies following pancreatectomy [8]. For this reason there is no information on the relationship between the disease genotype and PHHI β-cell function. In this context, recent work with the exon 35 mutation has provided novel insights into how the variations in the SUR1 gene can lead to the clinical manifestation of insulin hypersecretion.

Key observations were made of two children with familial PHHI who were both homozygous for a guanine to adenine point mutation in codon 1,437 of the SUR1 gene which in health encodes for an arginine residue. This mutation occurs at the 3′ end of exon 35 of SUR1 gene and results in the loss of a MSP1 restriction endonuclease recognition site. As a consequence exon 35 of the genomic DNA is no longer transcribed and the frame shift formed in this manner produces a premature stop codon 23 amino acids (69 base pairs) 3′ to the abnormally spliced exon 34, resulting in truncation of the SUR1 gene product by 200 amino acids. This loss includes the second nucleotide-binding fold of the SUR1 protein. The functional association of this mutation with defects in K_{ATP} channels has now established PHHI as a disorder arising from the "knock-out" of β-cell K_{ATP} channels [30]. Recombinant molecular biology techniques confirmed this hypothesis as a parallel mutation engineered in the SUR1 gene was unable to reconstitute K_{ATP} channels when coexpressed with wild-type Kir6.2 in a mammalian heterologous expression system [9].

An alternative approach to studying K_{ATP} channel function was taken by Miki et al. [31]. In this work transgenic mice were generated that expressed a dominant negative form of Kir6.2 in β-cells. This was achieved by substituting a glycine for serine at position 132 which is within the pore-forming region of the channel. Affected mice develop hyperinsulinaemic hypoglycaemia in vivo during the neonatal period, and β-cells isolated from the animals have: 1) impaired K_{ATP} channel function, 2) depolarised membrane potentials, and 3) elevated levels of cytosolic Ca^{2+}. These symptoms in mice therefore mimic PHHI in humans. Interestingly, the studies of Miki et al. [31] also show that as the transgenic animals develop, the period of hyperinsulinaemia/hypoglycaemia is subsequently followed in older rodents by hyperglycaemia and hyperinsulinaemia, i.e., the onset of diabetes. One explanation for this is that the authors also report a high frequency of apoptotic β-cell death in the transgenic mice. These data suggest that the unregulated Ca^{2+} influx has a compound pathology, underpinning inappropriate insulin secretion *and* leading to the induction of premature apoptosis. There is good clinical support for this model. Some children with PHHI who respond to long-term medical therapy with somatostatin and therefore develop without the need for pancreatectomy, do, however, remain refractory to standard oral glucose tolerance tests later in life [32]. This implies that through β-cell apoptosis, symptoms of the disease are alleviated yet there is still impaired β-cell function, and presumably decreases in β-cell numbers.

β-cell defects and medical therapy

It is perhaps surprising that given the loss of K_{ATP} function and PHHI, that the mainstay of medical therapy for this condition is the administration of diazoxide and somatostatin: hyperpolarizing agents of β-cells that activate K_{ATP} channels under normal conditions.

At least part of this discrepancy can be explained, certainly in the case of somatostatin, by non-K_{ATP} channel mediated events: inhibition of Ca^{2+} channels, modulation of adenylate cyclase and the termination of "distal signalling" (for a more detailed review see [33]). However, it has been shown that somatostatin and diazoxide terminate spontaneous β-cell electrical activity in PHHI tissue [10]. In these experiments it was shown that despite the absence of K_{ATP} channels the activation of a "novel" K^+ channel led to a hyperpolarization of the membrane and the termination of voltage-gated action potentials. This novel channel designated "K_{PHHI}" is K^+-selective and nucleotide-regulated. However, it is not clear how SUR1 and/or Kir6.2 gene defects relate to the function or the expression of these channels in β-cells, or indeed what the function of these channels is in control β-cell. Despite this, pharmacological regulation of the channels by diazoxide and by somatostatin readily explains why some patients, despite the absence of K_{ATP} channels are responsive to medical therapy.

Through advances in understanding the β-cell pathology in PHHI, it has also been possible to suggest "alternative" approaches to medical therapy for the condition. One key example of this is the use of voltage-gated Ca^{2+} channel blockers, which will terminate Ca^{2+} action potentials in vitro and alleviate hypoglycaemia in PHHI patients in vivo [7]. Critically, these findings also serve to authenticate the developing model of PHHI β-cells and the sequence of events that ensue following the loss of K_{ATP} channel function (Fig. 3).

The molecular events of insulin release

Capacitance measurement [34,35] is another example of the versatility of the patch-clamp technique and is used to monitor the final events in the cascade of signals that culminate in the release of insulin, i.e., exocytosis. The technique is based upon the observation that the outer membrane of a cell has a specific capacitance of approximately 10 μF/mm^2. As the process of exocytosis involves the fusion of the internal secretory granule membrane with the plasma membrane, this will result in an increase in cell surface area, and secretion can therefore be monitored as a parallel increase in cell capacitance. Additionally, capacitance measurements can also be used to study the reverse process, i.e., endocytosis, when the excess membrane is reclaimed, thereby producing a decrease in cell capacitance [36]. This powerful technique allows the process of exocytosis to be investigated in individual cells with a high temporal and spatial resolution. As capacitance measurements monitor secretion from the individual cell, the method also has the advantage of being able to distinguish between mechanisms regulating insulin secretion on the subcellular level and the events that involve changes in the number of cells contributing to a response.

The ability to assess secretion from individual cells has provided key insights into the normal regulatory mechanism of insulin release by examining the role of calcium ions, cAMP, protein kinases, G-proteins, and sulphonylureas [36–40]. Since PHHI is a disorder of insulin-secreting cells related to SUR1 gene mutations, the application of these procedures will be invaluable in elucidating the relationship between K_{ATP} channel genetic variations, secretion and the control of exocytosis/endocytosis. These studies may reveal new target sites for the development of novel agents for PHHI and diabetes therapy.

Concluding remarks

Studies of this type, in which the modus operandi is to integrate genetic information with molecular and cellular detail in relation to the understanding of a rare disease, clearly demonstrate how key lessons in biomedical science can be learned. The immediate impact of these findings is important to the clinical management of a disorder, but in the wider context the work also enriches our knowledge of cellular control mechanisms. In this paper, we have documented how the fundamental properties of insulin-secreting cells are compromised by genetic variations that impair ion channel activities. These events contrive to cause a potentially lethal disease, for which medical therapy is only of very limited use. Work in this field has provided a greater understanding of the neonatal disease PHHI, and this will hopefully lead to advances in the development of new drug therapies or strategies to alleviate the condition. With specific gene defects linked to the disorder, there is also the possibility that gene therapy techniques could one day be of value. From the perspective of ion channel function, these studies have demonstrated that K_{ATP} channels not only have a fundamental role to play in the ordered control of insulin release, but that they are also critical for β-cell survival.

Acknowledgements

Work in our laboratories is supported by the British Diabetic Association, the Medical Research Council Studentship (MJD), and by Stiftelsen Clas Groschinskys minnesfond, Tore Nilssons Stiftelse, Åke Wibergs Stiftelse, Wilhelm och Martina Lundgrens vetenskapsfond, Swedish Medical Research Council (CÄ).

References

1. Dean PM, Matthews EK. Electrical activity in pancreatic islet cells. Nature 1968;219:389–390.
2. Matthews EK, Sakamoto Y. Electrical characteristics of pancreatic islet cells. J Physiol 1975; 246:421–437.
3. Cook DL, Hales CN. Intracellular ATP directly blocks K^+ channels in pancreatic β-cells. Nature 1984;311:271–273.
4. Ashcroft F, Harrison DE, Ashcroft SJH. Glucose induces closure of single potassium channels in isolated rat pancreatic islets. Nature 1984;312:446–448.
5. Inagaki N, Gonoi T, Clement JP IV, Namba N, Inazawa J, Gonzalez G, Aguilar-Bryan L, Seino S, Bryan J. Reconstitution of I_{KATP}: an inward rectifier subunit plus the sulphonylurea receptor. Science 1995;270:1166–1170.
6. Sakura H, Ämmälä C, Smith PA, Gribble FM, Ashcroft FM. Cloning and functional expression of the cDNA encoding a novel ATP-sensitive potassium channel expressed in pancreatic β-cells, brain, heart and skeletal muscle. FEBS Lett 1995;377:338–344.
7. Lindley KJ, Dunne MJ, Kane C, Shepherd RM, Squires PE, James RFL, Johnson PRV, Eckhart S, Wakeling E, Dattani M, Milla PJ, Aynsley-Green A. Ionic control of β-cells function in nesidioblastosis. A possible therapeutic role for calcium channel blockade. Arch Dis Child 1996; 74:373–378.
8. Kane C, Shepherd RM, Squires PE, Johnson PRV, James RFL, Milla PJ, Aynsley-Green A, Lindley KJ, Dunne MJ. Loss of functional K_{ATP} channels in β-cells causes persistent hyperinsulinaemic hypoglycaemia of infancy. Nature Med 1996;2:1344–1347.
9. Dunne MJ, Kane C, Shepherd RM, Sanchez JA, James RFL, Johnson PRV, Aynsley-Green A, Lu S, Clement JP IV, Lindley KJ, Seino S, Aguilar-Bryan L. Familial persistent hyperinsulinemic hypoglycaemia of infancy and mutations in the sulfonylurea receptor. N Engl J Med 1997; 336:703–706.
10. Kane C, Lindley KJ, Johnson PRV, James RFL, Milla PJ, Aynsley-Green A, Dunne MJ. Therapy for persistent hyperinsulinaemic hypoglycaemia of infancy (PHHI); understanding the responsiveness of β-cells to diazoxide and somatostatin. J Clin Invest 1997;100:1888–1893.
11. Clement JP IV, Kunjilwar K, Gonzalez G, Schwanstecher M, Panten U, Aguilar-Bryan L, Bryan J. Association and stoichiometry of K(ATP) channel subunits. Neuron 1997;18:827–838.
12. Shyng S, Nichols CG. Octameric stoichiometry of the K_{ATP} channel complex. J Gen Physiol 1997;110:655–664.
13. Shyng S, Ferrigni T, Nichols CG. Control of rectification and gating of cloned K_{ATP} channels by the Kir6.2 subunit. J Gen Physiol 1997;110:141–153.
14. Shyng S, Ferrigni T, Nichols CG. Regulation of K_{ATP} channel activity by diazoxide and MgADP. Distinct functions of the two nucleotide binding folds of the sulfonylurea receptor. J Gen Physiol 1997;110:643–654.
15. Gribble FM, Tucker SJ, Ashcroft FM. The interaction of nucleotides with the tolbutamide block of cloned ATP-sensitive K^+ channel currents expressed in *Xenopus* oocytes: a reinterpretation. J Physiol 1997;504:35–45.
16. Nichols CG, Shyng SL, Nestorowicz A, Glaser B, Clement JP IV, Gonzalez G, Aguilar-Bryan L, Permutt MA, Bryan J. Adenosine diphosphate as an intracellular regulator of insulin secretion.

Science 1996;272:1785–1787.

17. Trapp S, Tucker SJ, Ashcroft FM. Activation and inhibition of K-ATP currents by guanine nucleotides is mediated by different channel subunits. Proc Natl Acad Sci USA 1997;94: 8872–8877.
18. Tucker SJ, Gribble FM, Zhao C, Trapp S, Ashcroft FM. Truncation of Kir6.2 produces ATP-sensitive K^+ channels in the absence of the sulphonylurea receptor. Nature 1997;387:179–183.
19. Ueda K, Inagaki N, Seino S. MgADP antagonism to Mg^{2+}-independent ATP binding of the sulfonylurea receptor SUR1. J Biol Chem 1997;272:22983–22986.
20. Thomas PM, Cote CJ, Wohlik N, Haddad B, Mathew PM, Rabl W, Aguilar-Bryan A, Gagel RF, Bryan J. Mutations in the sulphonylurea receptor gene in familial hyperinsulinemic hypoglycaemia of infancy. Science 1995;268:426–429.
21. Glaser B, Chiu KC, Anker R, Nestorowicz A, Landau H, Ben-Bassat H, Sholmai Z, Kaiser N, Thornton PS, Stanley CA, Spielman RS, Gogolin-Ewens K, Cerasi E, Baker L, Rice J, Donis-Keller H, Permutt, MA. Familial hyperinsulinism maps to chromosome 11p14–15.1, 30 cM centromeric to the insulin gene. Nature Genet 1994;7:185–188.
22. Aynsley-Green A, Polak JM, Bloom SR, Gough MH, Keeling J, Ashcroft SJH, Turner RC, Baum RC. Nesidioblastosis of the pancreas: definition of the syndrome and the management of the severe neonatal hyperinsulinaemic hypoglycaemia. Arch Dis Child 1981;56:496–508.
23. Ämmälä C, Cosgrove KE, James RFL, Aynsley-Green A, Lindley KJ, Dunne MJ. Potassium currents in normal and disease human β-cells. Diabetologica 1997;40:S1, A13.
24. Thomas P, Yuyang Y, Lightner E. Mutation of the pancreatic islet inward rectifier KIR6.2 also leads to familial persistent hyperinsulinaemic hypoglycaemia of infancy. Hum Mol Genet 1996;5:1809–1812.
25. Nestorowicz A, Inagaki N, Gonoi T, Schoor KP, Wilson BA, Glaser B, Landau H, Stanley CA, Thornton PS, Seino S, Permutt MA. A nonsense mutation in the inward rectifier potassium channel gene, Kir6.2, is associated with familial hyperinsulinism. Diabetes 1997;46:1743–1748.
26. Thomas PM, Cote GJ, Wohlik N, Haddad B, Mathew PM, Rabl W, Aguilar-Bryan L, Gagel RF, Bryan J. Mutations in the sulphonylurea receptor gene in familial persistent hyperinsulinaemic hypoglycaemia of infancy. Science 1995;268:426–429.
27. Thomas PM, Wohlik N, Huang E, Kuhnle U, Rabl W, Gagel RF, Cote GJ. Inactivation of the first nucleotide binding fold of the sulphonylurea receptor and familial persistent hyperinsulinaemic hypoglycaemia of infancy. Am J Hum Genet 1996;59:510–518.
28. Nestorowicz A, Wilson BA, Schoor KP, Inoue H, Glaser B, Landau H, Stanley CA, Thornton PS, Clement JP, Bryan J, Aguilar-Bryan L, Permutt MA. Mutations in the sulphonylurea receptor gene are associated with familial hyperinsulinism in Ashkenazi Jews. Hum Mol Genet 1996;5:1813–1822.
29. Nichols CG, Shyng SL, Nestorowicz A, Glaser B, Clement JP, Gonzalez G, Aguilar-Bryan L, Permutt MA, Bryan J. Adenosine-diphosphate as an intracellular regulator of insulin secretion. Science 1996;272:1785–1787.
30. Dunne MJ, Aynsley-Green A, Lindley KJ. Nature's K_{ATP} channel knock out; PHHI a K^+ channel disorder of pancreatic β-cells leading to hypersecretion of insulin. NIPS 1997;12:197–203.
31. Miki T, Tashiro F, Iwanaga T, Nagashima K, Yoshitomi H, Aihara H, Nitta Y, Gonoi T, Inagaki N, Miyazaki JI, Seino S. Abnormalities of pancreatic islets by targeted expression of a dominant-negative KATP channel. Proc Natl Acad Sci USA 1997;94:11969–11973.
32. Leibowitz G, Glaser B, Higazi AA, Salameh M, Cerasi E, Landau H. Hyperinsulinaemic hypoglycaemia of infancy (nesidioblastosis) in clinical remission: high incidence of diabetes mellitus and persistent beta cell dysfunction at long term follow up. J Clin Endocrinol Metab 1995;7: 185–188.
33. Dunne MJ, Ämmälä C, Straub SG, Sharp GWG. (1998). The electrophysiology of the β-cell and mechanisms of inhibition of insulin release. In: Jefferson LS, Cherrington AS (eds) The Handbook of Physiology, 1998 Edition (In press).

34. Neher E, Marty A. Discrete changes in of cell membrane capacitance observed under conditions of enhanced secretion in bovine adrenal chromaffin cells. Proc Natl Acad Sci USA 1982;72: 6712–6716.
35. Lindau M, Neher E. Patch-clamp techniques for time-resolved capacitance measurements in single cells. Pflügers Arch 1988;411:137–146.
36. Eliasson L, Proks P, Ämmälä C, Ashcroft FM, Bokvist K, Renström E, Rorsman P, Smith PA. Endocytosis of secretory granules in mouse pancreatic β-cells evoked by transient elevation of cytosolic calcium. J Physiol 1996;493:755–767.
37. Ämmälä C, Ashcroft FM, Rorsman P. Calcium-independent potentiation of insulin release by cyclic AMP in single β-cells. Nature 1993;363:356–358.
38. Ämmälä C, Eliasson L, Bokvist, K, Berggren P-O, Honkanen RE, Sjöholm A, Rorsman P. Activation of protein kinases and inhibition of protein phosphatases play a central role in the regulation of exocytosis in mouse pancreatic beta cells. Proc Natl Acad Sci USA 1994;91:4343–4347.
39. Bokvist K, Eliasson L, Ämmälä C, Renström E, Rorsman P. Co-localization of L-type Ca^{2+} channels and insulin-containing secretory granules and its significance for the initiation of exocytosis in mouse pancreatic β-cells. EMBO J 1995;14:50–57.
40. Eliasson L, Renström E, Ämmälä C, Berggren P-O, Bertorello AM, Bokvist K, Chibalin A, Deeney JT, Flatt PR, Gäbel J, Gromada J, Larsson O, Lindström P, Rhodes CJ, Rorsman P. PKC-dependent stimulation of exocytosis by sulfonylureas in pancreatic beta cells. Science 1996;271:813–815.

Dynamic Cells: Cell Biology of the 21st Century
S. Yagihashi, T. Kachi and M. Wakui, editors.

Calcium mobilization by glucagon-like peptide-I in pancreatic β-cells

Makoto Wakui, Sechiko Suga, Takahiro Kanno and Noritaka Kamimura
Department of Physiology, Hirosaki University School of Medicine, Hirosaki, Japan

Keywords: cAMP, L-type Ca^{2+} channel,protein kinase A.

Glucagon-like peptide-I (GLP-I)

The oral ingestion of glucose caused a much higher insulin secretion than the intravenous injection of glucose, although these two procedures resulted in nearly identical plasma glucose levels [1,2]. These findings made us seek a possible connection between the gastrointestinal tract and the endocrine pancreas. To date, it has been cleared that the oral injection of glucose induces the secretion of insulin by stimulating pancreatic β-cells, not only with an elevation of the plasma level of glucose, but also with gastrointestinal hormones such as GLP-I and glucose-dependent insulin-secreting polypeptide (GIP) released into the blood from intestinal mucosa [3]. Molecular cloning techniques have revealed that GLP-I as well as GIP are peptide hormones belonging to the glucagon-secretin-vasoactive intestinal polypeptide (VIP) family [4—7]. All of the members of the family affect various physiological functions including the secretion of insulin via activation of the cAMP/protein kinase A (PKA) signaling pathway. Among the members of the glucagon-secretin-VIP family of peptide hormones, GLP-I seems to be of particular importance for controlling insulin secretion and glucose metabolism [3].

GLP-I is synthesized mainly in intestinal L-cells [8,9]. From the preprogluca-gon of 180-amino acid compound, L-cells can produce glicentin (proglucagon (1—69)), and GLPs (72—158). The GLPs are secreted from the L-cells to the extracellular space in response to meal ingestion [10], and in turn enter the circulation. The GLPs include GLP-I (78—107) and GLP-II (126—158). Among these two GLPs, the only truncated forms of GLP-I show biological activities. GLP-I is further classified into two forms, GLP-I (7—37) and GLP-I (7—36) amide. GLP-I (7—36) amide is the predominant natural form of the peptide [11]. The basal circulating levels of GLP-I vary between 10 and 100 pM [12], and show 6- to 8-fold increase after the oral glucose ingestion [13]. The stimulatory effect of GLP-I on glucose-induced insulin secretion was first demonstrated in the per-

Address for correspondence: Makoto Wakui, Department of Physiology, Hirosaki University School of Medicine, Hirosaki 036, Japan.

fused rat pancreas [14–16]. In those studies, the stimulatory effect of GLP-I was shown to be dependent on the glucose concentration in the perfusate.

Although the activation of the cAMP/PKA signaling pathway is one of the major mechanisms by which GLP-I stimulates the secretion of insulin from pancreatic β-cells [17,18], recent investigations suggest that GLP-I and other members of the glucagon-secretin-VIP family also cause an elevation of intracellular concentration of Ca^{2+} ($[Ca^{2+}]_i$) in the pancreatic β-cells [19–22]. The increase in $[Ca^{2+}]_i$ also leads to potentiation of the secretion of insulin [23]. Interestingly, the elevation of $[Ca^{2+}]_i$ in response to GLP-I depends also on the extracellular concentration of glucose [20,24]. Here, we have focused our attention on the mechanism of Ca^{2+} mobilization by GLP-I in pancreatic β-cells.

GLP-I receptor and its signal transduction

Like other hormones of the glucagon-secretin-VIP family, GLP-I performs its physiological action on the pancreatic β-cells (potentiation of insulin secretion) through binding to its own receptor present in the cell membrane. The specific receptors for GLP-I have been detected on several insulinoma cell lines [25–27] and the normal rat pancreatic β-cells [28]. The primary structure of GLP-I receptor has been revealed by the cDNA cloning technique [29–31]. The GLP-I receptor consists of 463 amino acids, and like other members of glucagon-secretin-VIP family, contains seven transmembrane domains, typical of the superfamily of heterotrimeric GTP-binding protein (G-protein)-coupled receptors. However, the structure of the receptors of glucagon-secretin-VIP family is largely different from that of other members of the superfamily of G-protein-coupled receptors. One particular feature of the receptors of the members of glucagon-secretin-VIP family is the presence of a long (120 amino acids) amino terminal extracellular domain. The K_d for binding of GLP-I to its receptor is 0.2–0.6 nM [17,18], values not far from those of plasma concentrations of GLP-I.

The GLP-I receptor is functionally coupled to a Gs-protein, the activation of which leads to activation of adenylate cyclase (AC). Therefore, GLP-I, as a consequence of activation of its receptor, causes an increase in the cytoplasmic level of cAMP synthesized from ATP by the active AC. The cAMP molecule activates the cAMP-dependent protein kinase (PKA). Through the phosphorylation of the critical proteins by PKA, GLP-I conducts not only the activation of gene transcription for synthesis of insulin [17,27], but also the potentiation of the secretion of insulin. Recent studies show further the precise mechanism of the cAMP/PKA-dependent signaling activated by GIP [32,33] and GLP-I [33] to potentiate exocytosis of insulin granules.

Ca^{2+} is a key intracellular signal for secretion of insulin

Insulin secreted into the blood is a key molecule in controlling the glucose homeostasis, and glucose in the plasma (extracellular fluid) is a key molecule in controlling the secretion of insulin from pancreatic β-cells. The insulin secretion by glucose is brought about by its Ca^{2+} mobilizing effect and a subsequent increase in $[Ca^{2+}]_i$ in the pancreatic β-cells [34]. Glucose triggers the Ca^{2+} mobilization in pancreatic β-cells by changing the electrical properties of the cell membranes.

The resting membrane potential of the β-cell is approximately −65 mV in the presence of glucose in the extracellular solution at relatively low concentrations (2–5 mM) [35]. This value of the resting potential is mainly due to opening of K^+ channels present in the membrane. There are various types of K^+ channels present in the membrane of the β-cell, and at the normal resting membrane potential, only K^+ channels opening are ATP-sensitive K^+ channels (K_{ATP}) [36]. In other words, the opening of K_{ATP} largely contributes to the development of the resting membrane potential in pancreatic β-cells. It must be mentioned, however, that the resting potential of the β-cell does not come only from the conductance of the membrane to K^+ (equilibrium potential, E_K, is about −90 mV), but also from a combination of conductances to K^+ and other types of ions (equilibrium potentials, less negative than E_K). Therefore, either the closure of K_{ATP} or the opening of channels other than K^+ channels causes depolarization of the β-cell. When the extracellular concentration of glucose is increased, a glucose transporter (GLUT 2), present in the β-cell membrane, allows a large amount of glucose to enter the β-cell. Metabolism of glucose taken by the β-cell leads to production of ATP. An elevation in the intracellular concentration of ATP and/or an elevation of the ATP:ADP ratio causes closure of K_{ATP}, which leads to development of depolarization of the β-cell [37,38]. The depolarization of the β-cell to the critical level conducts opening of voltage-dependent Ca^{2+} channels, which allows extracellular Ca^{2+} to enter the cell, resulting in an elevation of $[Ca^{2+}]_i$.

As described above, the voltage-dependent Ca^{2+} channels play an important role in Ca^{2+} entry and subsequent increase in $[Ca^{2+}]_i$ in the pancreatic β-cells. Two types of voltage-dependent Ca^{2+} channels have been found in the β-cell membrane; one is characteristic of the L-type channel, and the other, of the T-type channel [36,39–41]. The physiological significance of the T-type Ca^{2+} channel is not clear, because the channel should be inactive in its state at the normal resting membrane potential of β-cells. On the other hand, the L-type Ca^{2+} channel is closed in its state at the resting potential and opened by depolarization of the membrane to the values more positive than −40 mV. Accordingly, the L-type Ca^{2+} channels play a particularly important role in regulation of the Ca^{2+} entry. The elevation of $[Ca^{2+}]_i$ potentiates the secretion of insulin by facilitating the docking of granules with the plasma membrane probably through phosphorylation of critical proteins by Ca^{2+}-dependent protein kinases [42,43].

Although it is not the case of glucose stimulation, the Ca^{2+} mobilization and subsequent increase in $[Ca^{2+}]_i$ occur through another pathway, the release of Ca^{2+} from intracellular stores such as endoplasmic reticulum (ER). The Ca^{2+} release from ER is triggered by an intracellular messenger, inositol triphosphate

($InsP_3$) [44] as is the case of a variety of other types of cells [45]. $InsP_3$ is formed by hydrolysis of PIP_2 by phospholipase C (PLC). For instance, acetylcholine, a transmitter released from parasympathetic nerve terminals, activates PLC through binding to its receptor coupled to the GTP-binding protein (Gq).

Ca^{2+} mobilization by GLP-I in pancreatic β-cells

As mentioned already, GLP-I is also known to cause an increase in $[Ca^{2+}]_i$. Through both the increase in $[Ca^{2+}]_i$ and the elevation of cytosolic levels of cAMP, GLP-I appears to potentiate insulin secretion from the β-cells. The increase in $[Ca^{2+}]_i$ induced by GLP-I is dependent on the extracellular glucose concentration [19—22]. Since the elevation of $[Ca^{2+}]_i$ induced by GLP-I is inhibited by application of dihydropyridines [20], the Ca^{2+} mobilization by GLP-I in pancreatic β-cells is likely to be due to Ca^{2+} entry through L-type Ca^{2+} channels. The gating of the L-type Ca^{2+} channel is voltage-dependent; therefore, GLP-I may induce either or both of the following two phenomena: one, depolarization of the membrane, and the other, potentiation of the voltage-sensitivity of L-type Ca^{2+} channels in β-cells.

GLP-I depolarizes the pancreatic β-cells

In the presence of glucose in the extracellular solution, GLP-I was shown to cause depolarization in the pancreatic β-cells [21,22,33]. However, neither the ionic mechanism nor the signal transduction responsible for the depolarization induced by GLP-I is clear at present. In our study on isolated rat pancreatic β-cells [46], GLP-I normally caused depolarization accompanied by a decrease in the membrane conductance. Some other β-cells, however, showed depolarization in response to GLP-I associated with an increase in the membrane conductance. One possible mechanism of the depolarization associated with a decrease in the membrane conductance is an inhibition of the K_{ATP} [21,22]. As mentioned already, the K_{ATP} is open at the resting state of the pancreatic β-cell, and its closure leads to development of membrane depolarization. Furthermore, in a possible inhibitory effect of GLP-I on the K_{ATP}, the activation of cAMP/PKA-signaling pathway seems to be involved, supported by a finding that GIP and glucagon, other cAMP-producing peptides, showed similar effect on the membrane potential to that of GLP-I [33]. However, evidence of the blocking of K_{ATP} by GLP-I has been not given yet, rather, it has been reported that GLP-I did not affect the K_{ATP} [47]. There is, furthermore, no evidence that the cAMP-dependent mechanism causes closure of the K_{ATP}. The presence of a synergistic effect of GLP-I and glucose on closing the K_{ATP} of β-cells was also shown, i.e., in the absence of extracellular glucose, GLP-I alone did not alter the membrane potential of the β-cell, but potentiated the magnitude of depolarization in response to 10 mM glucose [21], indicating that GLP-I accelerated the closure of K_{ATP} by glucose.

On the other hand, the cAMP-dependent mechanism activated by GLP-I was also shown to increase the conductance of nonselective cationic channels in pancreatic β-cells, which can also be responsible for the development of depolarization [48].

GLP-I activates L-type Ca^{2+} channels of β-cells

Because of particular importance of the L-type Ca^{2+} channels as a route of Ca^{2+} entry in pancreatic β-cells, it is valuable to know whether GLP-I acts directly (through other processes than developing depolarization) on the particular channels. It has been reported that GLP-I slows the inactivation kinetics of gating of the L-type Ca^{2+} channel of pancreatic β-cells, which allows a larger amount of Ca^{2+} to enter the cell [47]. This effect of GLP-I was mediated by the cAMP-dependent mechanism. In fact, application of the membrane-permeable cAMP analogue or an inhibitor of phosphodiesterase to the β-cell clearly slowed the time course of decay of the currents through L-type Ca^{2+} channels, showing little effect on the amplitude of the Ca^{2+} currents [49,50]. Therefore, the principal action of cAMP-dependent mechanism on the Ca^{2+} current in the β-cell was considered to be slowing the inactivation rather than increasing the maximal conductance of the channel gating [49].

However, there is evidence that cAMP and agents that raise the cytosolic cAMP level clearly augment the amplitude of glucose-dependent islet-cell bursting (Ca^{2+}-dependent action potentials) [51], indicating the cAMP-dependent mechanism to increase the maximal conductance of the membrane to Ca^{2+}. Furthermore, in many other types of cells including cardiac muscle [52] and skeletal muscle [53], the cAMP-dependent mechanism is shown to augment the Ca^{2+} currents through L-type Ca^{2+} channels. Our recent studies also indicate that GLP-I acts positively on the excitation kinetics of the gating of L-type Ca^{2+} channels in pancreatic β-cells. In our study on isolated rat pancreatic β-cells [46], GLP-I showed a concentration-dependent augmentation of the peak magnitude of Ba^{2+} currents through L-type Ca^{2+} channels. In addition, GLP-I shifted the I—V curve for the Ba^{2+} currents to the left (toward more negative). The half-activation potential (V1/2) shifted toward more negative value with GLP-I. Since all of these excitatory effects of GLP-I on L-type Ca^{2+} channels were not observed in β-cells pretreated with Rp-cAMPS, an inhibitor of PKA, GLP-I is likely to activate the L-type Ca^{2+} channel in a cAMP-dependent mechanism. Similar excitatory effects on L-type Ca^{2+} channels were observed with application of dibutyryl cAMP [54]. The increase in the voltage sensitivity of the L-type Ca^{2+} channel in response to GLP-I allows the Ca^{2+} channel to be excited by an even smaller depolarization, which also potentiates the glucose effect on Ca^{2+} entry and the subsequent secretion of insulin.

Glucose dependency of Ca^{2+} mobilizing effect of GLP-I

The secretion of insulin by GLP-I depends on extracellular concentration of glucose. One of the explanations for the glucose-dependency of GLP-I effect, is that the rate of cAMP production by activation of AC in response to GLP-I may require intracellular Ca^{2+}/calmodulin complex [55], and glucose is a factor to give an elevation of $[Ca^{2+}]_i$. On the other hand, Ca^{2+} mobilization by GLP-I is also extracellularly glucose-dependent as described already. Since the Ca^{2+} mobilization by GLP-I is due to Ca^{2+} entry via development of depolarization and/ or activation of L-type Ca^{2+} channel, it is worth exploring whether these two mechanisms show a glucose-dependency. The activation by GLP-I of L-type Ca^{2+} channel is mediated by activation of the cAMP signaling. Since cAMP is synthesized from ATP released by the metabolism of nutrients including glucose, the cytosolic level of ATP should affect the rate of production of cAMP. In this respect, excitatory GLP-I action on the L-type Ca^{2+} channel through the cAMP-dependent mechanism is indirectly dependent on the amount of glucose entering the cell.

On the other hand, depolarization induced by GLP-I also seems to be glucose-dependent, as shown by the finding that GLP-I accelerated the glucose-induced depolarization [21]. In our experiment, GLP-I also showed little effect on the membrane potential of the β-cell under a condition where the cytosolic level of ATP was nominally free [46]. Together, these findings indicate that the cytosolic ATP is needed for GLP-I to cause depolarization (probably to close K_{ATP}), and glucose is required for production of ATP. In addition, it should be also noted that the magnitude of the resting membrane potential depends on the cytosolic levels of ATP, i.e., the higher the concentration of cytosolic ATP, the lower (less negative) the magnitude of the resting potential. Therefore, in a condition where the resting membrane potential is low with a higher concentration of glucose, depolarization induced by GLP-I by any ionic mechanism easily reaches the threshold to open the L-type Ca^{2+} channels.

References

1. Elick H, Stimmler L, Hlad CJ Jr, Arai Y. Plasma insulin response to oral and intravenous glucose administration. J Clin Endocrinol 1964;24:1076–1082.
2. Perley MJ, Kipnis DM. Plasma insulin responses to oral and intravenous glucose: studies in normal and diabetic subjects. J Clin Invest 1967;46:1954–1962.
3. Fehmann HC, Goke R, Goke B. Cell and molecular biology of the incretin hormones glucagon-like peptide-I and glucose-dependent insulin releasing polypeptide. Endocrine Rev 1995;16: 390–410.
4. Bell GI, Santerre RF, Mullenbach GT. Hamster preproglucagon contains the sequence of glucagon and two related peptides. Nature 1983;302:716–718.
5. Heinrich G, Gros P, Habener JF. Glucagon gene sequence: four of six exons encode separate functional domain of rat preproglucagon. J Biol Chem 1984;259:14082–14087.
6. Takeda J, Seino Y, Tanaka K, Fukumoto H, Kayano T, Takahashi H, Mitani T, Kurono M, Suzuki T, Tobe T, Imura H. Sequence of an intestinal cDNA encoding human gastric inhibitory polypeptide precursor. Proc Natl Acad Sci USA 1984;84:7005–7008.
7. Inagaki N, Seina Y, Takeda J, Yano H, Yamada Y, Bell GI, Eddy RL, Fukushima Y, Byers MG, Shows TB, Imura H. Gastric inhibitory polypeptide: structure and chromosomal localization

of the human gene. Molec Endocrinol 1989;3:1014–1021.

8. Orskov C, Holst JJ, Knuhtsen S, Baldissera FGA, Poulsen SS, Vagn Nielsen O. Glucagon-like peptides GLP-1 and GLP-2, predicted products of the glucagon gene, are secreted separately from pig small intestine but not pancreas. Endocrinology 1986;119:1467–1475.
9. Mojsov S, Kopczynski MG, Habener JF. Both amidated and nonamidated forms of glucagon-like peptide I are synthesized in the rat intestine and the pancreas. J Biol Chem 1990;265: 8001–8008.
10. Creutzfeldt W, Nauck MA. Gut hormones and diabetes mellitus. Diabetes Metab Rev 1992; 8:149–177.
11. Orskov C, Bersani M, Johnsen AH, Hojrup P, Holst JJ. Complete sequences of glucagon-like peptide I from human and pig small intestine. J Biol Chem 1989;264:12826–12829.
12. Orskov C, Jeppesen J, Madsbad S, Holst JJ. Proglucagon products in plasma of noninsulin-dependent diabetics and nondiabetic controls in the fasting state and after oral glucose and intravenous arginine. J Clin Invest 1991;87:415–423.
13. Herrmann C, Goke R, Richter G, Fehmann HC, Arnold R, Goke B. Glucagon-like peptide-I and glucose-dependent insulin-releasing polypeptide secretion in response to nutrients. Digestion 1995;56:117–126.
14. Holst JJ, Orskov C, Vagan Nielsen O, Schwartz TW. Truncated glucagon-like peptide I, an insulin-releasing hormone from the distal gut. FEBS Lett 1987;211:167–174.
15. Mojsov S, Weir GC, Habener JF. Insulinotropin: glucagon-like peptide I(7-37) co-encoded in glucagon gene is a potent stimulator of insulin release in the perfused rat pancreas. J Clin Invest 1987;79:616–619.
16. Weir GC, Mojsov S, Hendrick GK, Habener JF. Glucagon-like peptide I(7-37) actions on endocrine pancreas. Diabetes 1989;38:338–342.
17. Drucker DJ, Philippe J, Mojsov S, Chick WL, Habener JF. Glucagon-like peptide I stimulates insulin gene expression and increases cyclic AMP levels in a rat islet cell line. Proc Natl Acad Sci USA 1987;84:3434–3438.
18. Goke R, Trautmann ME, Haus E, Richter G, Fehmann HC, Arnold R, Goke G. Signal transduction after GLP-I(7-36)amide binding in RINm5F cells. Am J Physiol 1989;257:G397–G401.
19. Lu M, Wheeler MB, Leng XH, Boyd AE III. The role of the free calcium level in β-cell signal transduction by gastric inhibitory polypeptide and glucagon-like peptide I(7-37). Endocrinology 1993;132:94–100.
20. Yada T, Itoh K, Nakada M. Glucagon-like peptide-i-(7-36)amide in adenosine 3′,5′-monophosphate increase cytosolic free Ca^{2+} in rat pancreatic β-cells by enhancing Ca^{2+} channel activity. Endocrinology 1993;133:1685–1692.
21. Holz IV GG, Kuhtreiber WM, Habener JF. Pancreatic beta-cells are rendered glucose-competent by the insulinotropic hormone glucagon-like peptide-1(7-37). Nature 1993;361:362–365.
22. Holz IV GG, Leech CA, Habener JF. Activation of a cAMP-regulated Ca^{2+}-signaling pathway in pancreatic β-cells by the insulinotropic hormone glucagon-like peptide-1. J Biol Chem 1995;270:17749–17757.
23. Wollheim CB, Sharp CWG. Regulation of insulin release by calcium. Physiol Rev 1981;61: 14–973.
24. Cullinan CA, Brady EJ, Saperstein R, Leibowitz MD. Glucose-dependent alterations of intracellular free calcium by glucagon-like peptide-I(7-36)amide in individual ob/ob mouse β-cells. Cell Calcium 1994;15:391–400.
25. Goke R, Conlon JM. Receptors for glucagon-like peptide 1(7-36)amide on rat insulinoma-derived cells. J Endocrinol 1988;116:357–362.
26. Orskov C, Nielsen JH. Truncated glucagon-like peptide-1 (proglucagon 78-107 amide), an intestinal insulin-releasing peptide, has specific receptors on rat insulinoma cells (RIN5AH). FEBS Lett 1988;229:175–178.
27. Fehmann HC, Habener JF. Insulinotropic hormone glucagon-like peptide-I(7-37) stimulation of

proinsulin gene expression and proinsulin biosynthesis in insulinoma βTC-1 cells. Endocrinology 1992;130:159–166.
28. Orskov C, Poulsen S. Glucagonlike peptide-1-(7-36)-amide receptors only in islets of Langerhans. Diabetes 1991;40:1292–1296.
29. Laburthe M, Couvineau A, Gaudin P, Maoret JJ, Rouyer-Fessard C, Nicole P. Receptors for VIP, PACAP, secretin, GFR, glucagon, GLP-1, and other members of their new family of G protein-linked receptors: structure-function relationship with special reference to the human VIP-1 receptor. Ann NY Acad Sci 1996;805:94–109.
30. Thorens B. Expression cloning of the pancreatic β-cell receptor for the gluco-incretin hormone glucagon-like peptide 1. Proc Natl Acad Sci USA 1992;89:8641–8645.
31. Dillon JS, Tanizawa Y, Wheeler MB, Leng XH, Ligon BB, Rabin DU, Yoo-Warren H, Permutt A, Boyd III AE. Cloning and functional expression of the human glucagon-like peptide-1 (GLP-1) receptor. Endocrinoligy 1993;133:1907–1910.
32. Ding WG, Gromada J. Protein kinase A-dependent stimulation of exocytosis in mouse pancreatic β-cells by glucose-dependent insulinotropic polypeptide. Diabetes 1997;46:615–621.
33. Gromada J, Bokvist K, Ding WG, Holst JJ, Nielsen HJ, Rorsman P. Glucagon-like peptide 1 (7-36)amide stimulates exocytosis in human pancreatic β-cells by both proximal and distal regulatory steps in stimulus-secretion coupling. Diabetes 1998;47:57–65.
34. Ashcroft FM, Ashcroft SJH. Mechanism of insulin secretion. In: Ashcroft FA, Ashcroft SJH (eds) Insulin: Molecular Biology to Pathology. New York: Oxford University Press, 1992; 97–150.
35. Henquin JC, Meissner HP. Significance of ionic fluxes and changes in membrane potential for stimulus-secretion coupling in pancreatic β-cells. Experientia 1984;40:1043–1052.
36. Ashcroft FM, Rorsman P. Electrophysiology of the pancreatic β-cell. Prog Biophys Molec Biol 1991;54:87–143.
37. Ashcroft FM, Harrison DE, Ashcroft S. Glucose induces closure of single potassium channels in isolated rat pancreatic β-cells. Nature 1984;312:446–448.
38. Petersen OH. Control of potassium channels in insulin-secreting cells. ISI Atras Sci 1988;1: 144–149.
39. Rorsman P, Ashcroft FM, Trube G. Single Ca channel currents in mouse pancreatic β-cells. Pflügers Arch 1988;412:597–603.
40. Satin LS, Cook DL. Evidence for two calcium currents in insulin-secreting cells. Pflügers Arch 1988;411:401–409.
41. Ashcroft FM, Kelly RP, Smith PA. Two types of Ca channel in rat pancreatic β-cells. Pflügers Arch 1990;415:504–506.
42. Matsumoto K, Fukunaga K, Miyazaki J, Shichiri M, Miyamoto E. Ca^{2+}/calmodulin-dependent protein kinase II and synapsin I-like protein in mouse insulinoma min 6 cells. Endocrinology 1995;136:3784–3793.
43. Renstrom E, Eliasson L, Bokvist K, Rorsman P. Cooling inhibits exocytosis in single mouse pancreatic β-cells by suppression of granule mobilization. J Physiol 1996;494:41–52.
44. Nilson T, Arkhammar P, Hallberg A, Fellman B, Berggren PO. Characterization of the inositol 1,4,5-triphosphate-induced Ca^{2+} release in pancreatic β-cells. Biochem J 1987;248:329–336.
45. Berridge MJ. Inositol triphosphate and calcium signaling. Nature 1993;361:315–325.
46. Suga S, Kanno T, Nakano K, Takeo T, Dobashi, Y, Wakui M. GLP-I(7-36)amide augments Ba^{2+} current through L-type Ca^{2+} channel of rat pancreatic β-cell in a cAMP-dependent manner. Diabetes 1997;46:1755–1760.
47. Britsch S, Drewa PK, Lang F, Gregor M, Drews G. Glucagon-like peptide-1 modulates Ca^{2+} current but not K^+ATP current in intact mouse pancreatic β-cells. Biochem Biophys Res Commun 1995;207:33–39.
48. Kato M, Ma HT, Tatemoto K. GLP-1 depolarizes the rat pancreatic β-cell in a Na^+-dependent manner. Reg Peptides 1996;62:23–27.
49. Ammala C, Ashcroft FM, Rorsman P. Calcium-independent potentiation of insulin release by

cyclic AMP in single β-cells. Nature 1993;363:356–358.
50. Gills KD, Misler S. Enhancers of cytosolic cAMP augment depolarization-induced exocytosis from pancreatic β-cells: evidence for effects distal to Ca^{2+} entry. Pflügers Arch 1993;424: 195–197.
51. Henquin JC, Meissner HP. Effects of theophylline and dibutyryl cyclic adenosine monophosphate on the membrane potential of mouse pancreatic β-cells. J Physiol 1984;351:595–612.
52. Trautwein W, Hescher J. Regulation of cardiac L-type calcium current by phosphorylation and G proteins. Ann Rev Physiol 1990;52:257–274.
53. Mundina-Weilenmann C, Ma J, Rios E, Hosey MM. Dihydropyridine-sensitive skeletal muscle Ca channels in polarized planar bilayers: effects of phosphorylation by cAMP-dependent protein kinase. Biophys J 1991;60:902–909.
54. Kanno T, Suga S, Wu J, Kimura M, Wakui M. Intracellular cAMP potentiates voltage-dependent activation of L-type Ca^{2+} channels in rat islet β-cells. Pflügers Arch 1998;435:578–580.
55. Valverd I, Vandermeers A, Anjaneyulu R, Malaisse WJ. Calmodulin activation of adenylate cyclase in pancreatic islets. Science 1979;206:225–227.

Dynamic Cells: Cell Biology of the 21st Century
S. Yagihashi, T. Kachi and M. Wakui, editors.

Impaired growth and premature death of islet β-cells in type 2 (non-insulin-dependent) diabetes mellitus

Soroku Yagihashi, Motoi Koyama, Ryu-ichi Wada, Hiroki Mizukami and Hiroyasu Sakuraba
Department of Pathology, Hirosaki University School of Medicine, Hirosaki, Japan

Abstract. Impaired development and premature death of islet β-cells have been invoked in the pathogenesis of diabetes mellitus. To explore the mechanism of β-cell loss in diabetes, we examined the islet pathology of Japanese type 2 (non-insulin-dependent) diabetic patients and their counterpart of the animal model (GK rat) with morphometric analysis. Marked reduction of islets with irregular contour and hyalinization was detected in diabetic patients and there was β-cell depletion to the extent of a 33% reduction compared to nondiabetic patients. GK rats also exhibited progressive loss of β-cells starting at 8 weeks of age, showing 25 and 50% reduction at 12 and 24 weeks of age, respectively. Young GK rats showed smaller replicative activity measured by bromo-deoxyuridine uptake and upregulated expression of Bcl-2 protein compared to control Wistar rats. Excessive hyperglycemia induced by 30% sucrose feeding for 8 weeks further facilitated the β-cell loss in GK rats, which was associated with the appearance of apoptotis and increased expression of 8-hydroxy-deoxyguanosine, suggesting the DNA damage of β-cells by oxidative stress. By contrast, α-glucosidase-inhibitor treatment for 24 weeks significantly inhibited the reduction of β-cells by 30%. These results indicated that both poor replicative capacity and increased oxidative stress are involved in the progressive depletion of β-cells in type 2 diabetes.

Keywords: apoptosis, Bcl-2, BrdU, GK rat, Japanese type 2 diabetic patient, morphometry.

Introduction

With drastic westernization of lifestyle, the populations of diabetic patients are now expanding in Asian countries including Japan, Korea and China. More than 95% of diabetic patients in these countries have non-insulin-dependent (type 2) diabetes mellitus. Impaired insulin secretion, rather than obesity eliciting insulin resistance, is a salient feature in oriental type-2 diabetes. By contrast, obesity with insulin resistance is characteristic in type 2 diabetic patients in western countries. Likewise, the islet pathology in type 2 diabetic patients is not uniform, varying from hyalinosis with depletion of endocrine cells to having no particular changes [1–4].

Diabetic animal models have served for the studies to clarify the pathogenesis of diabetes mellitus and to explore useful treatment modalities for the prevention and management of this common disorder. The process of islet cell destruction has been vigorously examined in spontaneously diabetic animal models with insu-

Address for correspondence: Dr S. Yagihashi, Department of Pathology, Hirosaki University School of Medicine, 5 Zaifu-cho, Hirosaki 036 Japan.

lin deficiencies. However, the islet pathology of type 2 diabetic animal models and the mechanism of the pathological processes has not been fully elucidated.

In this paper, the authors attempt to present characteristic pathology of pancreatic islets in Japanese type 2 diabetic patients and in GK rat, a spontaneously onset non-insulin-dependent diabetic rat, a suitable model for human type 2 diabetes [5,6]. The significance of impaired growth in the development of islet lesions will be discussed from the data obtained from studies on this animal model.

Islet pathology of Japanese type 2 diabetic patients

The pancreas was examined in the autopsy cases of Japanese type 2 diabetic patients. Nondiabetic subjects dying of cardiovascular or malignant neoplastic diseases were used for comparison. The weight of the diabetic pancreas was not significantly different from nondiabetic subjects. Fatty infiltration in the peripheral pancreatic parenchyma was common with ductulization of acinar tissues. There were interstitial fibrosis and vascular sclerotic changes in older diabetic subjects. Most conspicuous changes in the islets were irregular outlines and atrophy with loss of endocrine cells. There were some islets replaced by hyalinous fibrosis often identified as amyloid deposition by Congo-red staining as well as positive immunostaining with human amylin antibodies. Specific immunostaining for insulin disclosed depletion of islet β-cells and swollen endocrine cells showing hydropic changes. With the morphometric analysis, population of islet β-cells was found to be depleted in type 2 diabetic patients as compared to nondiabetic subjects (Fig. 1). In nondiabetic subjects, islet volume was 3.3% on average, compared to type 2 diabetic patients of 2.8%. Approximately 70% of islet mass was occupied by β-cell areas in nondiabetic subjects, whereas only 50% were so in type 2 diabetic patients. It is therefore concluded that Japanese diabetic patients do express a reduced population of islet β-cells, although to a lesser extent.

Islet changes in GK rat, a type 2 diabetic animal model

Type 2 diabetes is characterized by impaired insulin secretion and insulin resistance. GK rats are spontaneously diabetic rats with noninsulin dependence. This model is nonobese, showing hyperglycemia and complication-like syndromes [5,6]. It has been suggested that processes of insulin secretion from insulin sensitivity through KATP channel inactivation as well as energy utilization of mitochondrial enzymes (glycerol-phosphate shuttles) are impaired in this model [7–9].

On the pancreatic sections stained with hematoxylin and eosin, the islet of GK rats at 5–6 weeks of age did not show any difference from age-matched control Wistar rats. After 8–12 weeks of age, the islet became fibrotic with irregular outlines. Vacuolation, and degranulation of islet cells were also found. Thereafter,

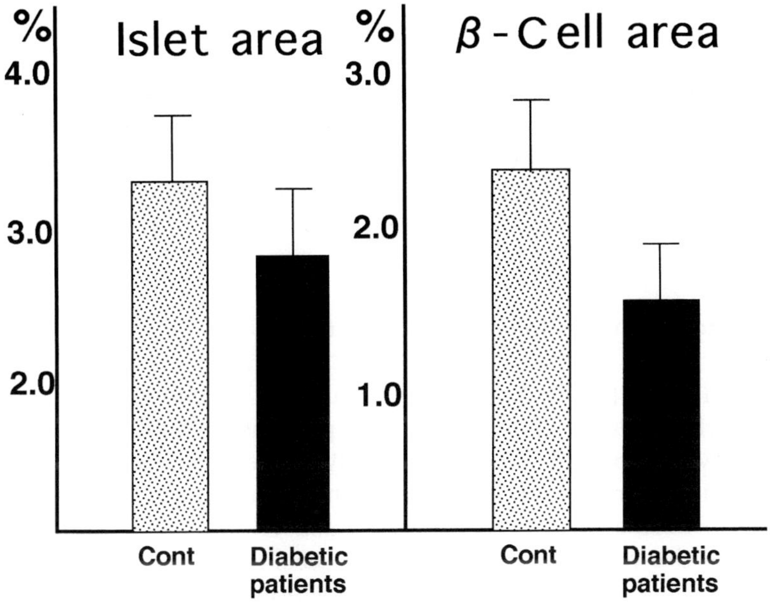

Fig. 1. Percentage area occupied in the pancreatic parenchyma by islets (left) and β-cells in Japanese type 2 diabetic patients (DM) and non-diabetic subjects (Cont).

there was progressive islet fibrosis and depletion of islet endocrine cells. Specific immunohistochemical staining with insulin antibodies confirmed that degeneration and loss of islet endocrine cells occurred mainly in β-cells. With morphometric analysis on insulin-stained slides, approximately 19% reduction of islet β-cell volume density was demonstrated in GK rats at 8 weeks of age. With increasing duration of diabetes, decrease in islet β-cell volume density became more severe, reaching about 49 and 68% reduction at 12 and 32 weeks of age, respectively. Similarly, there was significant reduction of islet β-cell mass in GK rats after 12 weeks of age compared to age-matched control Wistar rats.

By contrast, immunohistochemical stainings using antiglucagon and antisomatostatin antibodies did not reveal significant differences in the population and mass of α- and δ-cells between GK rats and age-matched control rats throughout the experimental period up to 36 weeks of age. In normal Wistar rats, there were no significant age-related changes in the islet cell population.

Proliferative activity of islet cells in GK rats

Proliferative activity of islet β-cells was examined by quantitation of positive cells which showed 5-bromo-2′-deoxy-uridine uptake (BrdU) in the nuclei within islets. For this purpose, BrdU (100 mg/kg) was injected intraperitoneally 4 h before sacrifice and then BrdU and insulin double immunostaining was conducted on the pancreatic paraffin sections using antibodies to insulin and BrdU.

The proliferative activity was expressed as a percentage of BrdU positive cells in 4,000 insulin-positive cells (BrdU Index).

In normal Wistar rats, the proliferative activity became smaller with aging. The mean index was about 2.5% at 5 weeks of age, whereas it was only 0.07% at 36 weeks of age in Wistar rats. The index of GK rats was significantly reduced by 30% at 5 weeks of age and by 45% of normal Wistar rats at 8 weeks of age. After 12 weeks of age, there was no significant difference in the BrdU labelings between GK rats and normal Wistar rats.

Bcl-2 and Bax immunoreactivities in islets of GK rat

To explore the mechanism of β-cell depletion, we examined immunohistochemically the expression of apoptosis-related oncogenic proteins of Bcl-2 and Bax. Throughout the observation period, islets were faintly stained with Bcl-2 in normal Wistar rats and there were no apparent alterations in the staining intensities with increasing age in normal Wistar rats. By contrast, GK rats at 6 weeks of age showed strong positive reactions to Bcl-2 in the islets compared to those in normal Wistar rats. However, these intensified reactions became less with increasing age in GK rats. There was no significant difference between GK rats and Wistar rats after 8 weeks of age, in which only weak reactions were noted in the islets of GK rats. On the other hand, islet cells were also positively stained with Bax in both GK and normal Wistar rats and the staining intensities did not alter with aging. There was no difference in the Bax staining reactions between GK and normal Wistar rats.

Effects of sucrose feeding on islets of GK rat

The effects of 30% sucrose feeding on the islet cell population were examined in GK rats of 6 weeks of age to explore if sucrose-induced hyperglycemic condition alters the islet cell population. The sucrose-fed GK rats showed more severe hyperglycemia at fast and after glucose load. With sucrose treatment for 6 weeks, islet changes became conspicuous showing early irregular contours and fibrotic changes. Morphometric analysis on the insulin-stained sections demonstrated that there was further reduction of mean β-cell volume density in GK rats, representing 52% of age-matched untreated GK rats (Fig. 2). It was indicated that continuous hyperglycemia accelerated reduction of islet β-cells.

The mechanism of sucrose-induced acceleration of β-cell loss was explored by the detection of apoptosis using an ApopTag in situ apoptosis detection kit. In normal Wistar rats, there were no apparent cells which underwent apoptosis. Without sucrose feeding, GK rats did not show apoptosis in the islets throughout the observation period up to 36 weeks of age. With sucrose feeding, however, we could demonstrate the presence of apoptosis in islets of GK rats from 0.067–0.056% levels of islet cell population. The apoptosis was found to occur only in the β-cells.

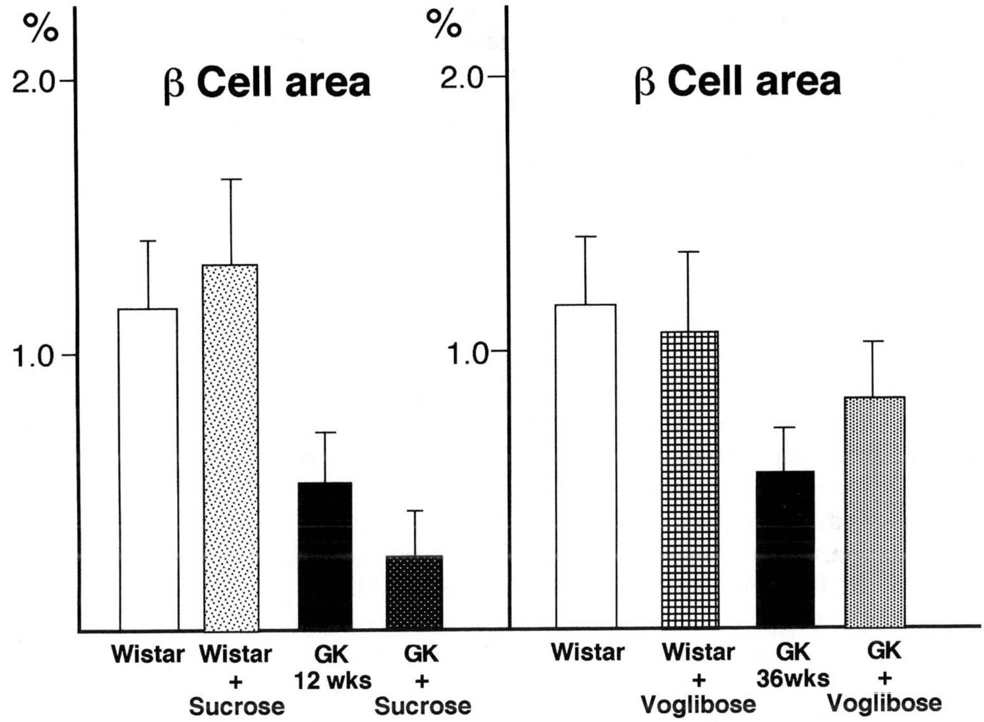

Fig. 2. Islet β-cell volume density in GK rats and age-matched normal Wistar rats under 30% sucrose feeding for 6 weeks and under treatment with α glucosidase inhibitor (Voglibose) for 24 weeks.

We then examined the intensity of oxidative stress represented by the expression of immunoreactive 8-hydroxy-deoxyguanosine (8-OH-dG) on the paraffin sections of experimental animals. There were weakly positive staining reactions of 8-OH-dG in ductal epithelial cells and smooth muscle cells of the ductal walls and islet cells in normal Wistar rats. Although the reactions were not uniform among islets, they appeared to be stronger in ductal cells and islet cells in GK rats than age-matched normal Wistar rats. The staining reactions were further intensified in islets of sucrose-fed GK rats, while those in normal Wistar rats did not alter after sucrose diet.

Effects of α-glucosidase inhibitor on islets of GK rat

Whether long-term blood glucose control can inhibit progressive loss of islet β-cells was examined in GK rats by treatment with α-glucosidase inhibitor (Voglibose; Takeda Pharmaceutical Ltd., Osaka) for 24 weeks. Elevated blood glucose levels at both fasting and nonfasting were significantly decreased in Voglibose-treated GK rats whose body weight was significantly reduced compared to untreated GK rats. Basal insulin levels and glycated hemoglobin levels were also decreased in treated GK rats.

Morphometric analysis on the insulin-stained pancreatic sections disclosed preservation of β-cells, inhibiting about 50% prevention of reduction of β-cell volume density in Voglibose-treated GK rats (Fig. 2). By contrast, there was no significant difference in BrdU labelings in β-cells between treated and untreated GK rats.

Discussion

The present study has shown that pancreatic islets in Japanese type 2 diabetic patients were characterized by a reduced number of islets with β-cell reduction, replaced by fibrosis and amyloid deposition. The percentage reduced in Japanese patients is in contrast with the data obtained from western type 2 diabetic patients which showed only a modest depletion of β-cell volume or population [2—4]. It is therefore apparent that the Japanese diabetic patients, although they were not commonly obese, exhibit a more severe degree of insulin deficiency at the end stage of the disorder. This may suggest that islets in oriental diabetic subjects are likely to be depleted with their population of β-cells, in part due to innate susceptibility to environmental insults or poor capacity for replication.

The salient physiological feature of type 2 diabetes is exemplified by an impaired insulin secretion of islet β-cells and insulin resistance. The relationship between insulin resistance and poor insulin secretion is not known and it is yet to be determined whether β-cells undergo degeneration as a consequence of insulin resistance. GK rats are known to be a good animal model for type 2 diabetes showing impaired insulin secretion and insulin resistance [5—7]. By our temporal observations on the pancreatic islets, the onset of diabetic syndrome is ascribed to progressive loss of islet β-cells, which was apparent at the age of 8 weeks in our GK rats of the Japanese colony. Even before the detection of significant loss of islet β-cells, GK rats expressed significantly smaller values of the proliferative activity of β-cells as estimated by BrdU labeling index, suggesting that the reduction of islet β-cells in GK rats is, at least in part, due to impaired proliferative activity during maturation.

The accelerated loss of β-cells in sucrose-fed GK rats suggested that continuous hyperglycemia promoted apoptotic-cell death of β-cells, leading to a smaller mass of islet β-cells. Our in situ detection of apoptotic cells, which was only possible in sucrose-fed GK rats, supports this contention, although the percentage of positive cells was small. By contrast, improvement of blood glucose control by treatment with α glucosidase inhibitor was found to inhibit the progressive depletion of β-cells. These findings provide evidence that a high glucose level is toxic to islet β-cells (glucotoxicity) determining the β-cell population in diabetic animal models. In GK rats, it may also be speculated that mild elevation of glucose levels have insidiously injured β-cells even before the age of 6 weeks together with innate insulin resistance, leading to significant levels of β-cell loss for glucose intolerance. The absence of significant effects of sucrose on the islet cell population in normal Wistar rats may also indicate that β-cells in GK rats are more susceptible to hyperglycemia.

Glucotoxicity is now mediated by several pathways, for removal or degradation of excessive glucose, represented by polyol pathway and nonenzymatic glycation. Cellular structural proteins as well as molecules including enzymes in the pancreatic islets are the preferential target for nonenzymatic glycation. Advanced glycation end-products (AGE) accumulate in the tissues containing long-lived proteins as a consequence of glycation, and a lot of oxidative stress is produced during the formation of AGE [10]. Such a glycation process is now implicated not only in the impairment of insulin secretion, but also in the apoptotic death of β-cells [11,12]. The glycation process is known to cleave the DNA to induce apoptosis of isolated islet β-cells [12]. Our observation, increased expression of 8-OH-dG in sucrose-fed GK rats, was in keeping with the data obtained from in vitro studies described above. It is not known, however, why only sucrose-fed GK rats show excessive oxidative stress in the islets but not sucrose-fed Wistar rats. One could consider that scavenger pathways for the oxidative-stress-like superoxide dismutase or reduced glutathione are relatively inactive in this model, serving susceptible condition of β-cells to free radicals.

Development and replication of islet β-cells are of paramount importance for the maintenance of energy metabolism in mammals [13]. Primitive precursor endocrine cells develop with the aid of growth factors like transforming growth factor, insulin-like growth factor I, activin A, and betacellulin [14—16]. During this process, principal regulating genes like insulin promoter factor, PDX-1 are considered to play an important role in the development of islet endocrine cells [16]. For the establishment of cell population of islet endocrine cells, a variety of the factors above possibly determine the islet cell population. Once the islets are established, the total population of islet cells is determined by the replicative capacity and death rate of the cells. The progressive reduction of islet β-cells detected in the GK rats is, therefore, ascribed to poor replicative capacity and premature death (apoptosis) of β-cells [17]. Bcl-2 protein is an oncogenic protein which acts against apoptotic death [18]. Enhanced expression of Bcl-2, but not of Bax, in young GK rats may, therefore be a reactive upregulation against the apoptotic process. It would be of interest to examine whether the growth factors are impaired in GK rats.

In conclusion, islets of both Japanese type 2 diabetic patients and GK rats are depleted with β-cells, which may largely be ascribed to innate poor proliferative activity and apoptotic death of islet β-cells. The susceptibility of islet cells to oxidative stress may be implicated in the mechanism of apoptosis.

Acknowledgements

We are greatly indebted to Drs Hitoshi Ikeda and Hiroyuki Odaka in Takeda Research Institute in Osaka for providing GK rats and to Ms Michiko Sugo, Yuko Sasaki, and Yaeko Tsushima for their technical assistance.

References

1. Saito K, Yaginuma N, Takahashi T. Differential volumetry of A, B, and D cells in the pancreatic islets of diabetic and nondiabetic subjects. Tohoku J Exp Med 1979;129:273–283.
2. Stefan Y, Orci L, Malaisse-Lagae F, Perrelet A, Patel Y, Unger RH. Quantitation of endocrine cell content in the pancreas of nondiabetic and diabetic humans. Diabetes 1982;31:694–699.
3. Rahier J, Goebbels RM, Henquin JC. Cellular composition of the human diabetic pancreas. Diabetologia 1983;24:366–371.
4. Clark A, Wells CA, Buley ID, Cruickshank JK, Vanhegan RI, Matthews DR, Cooper GJS, Holman RR, Turner RC. Islet amyloid, increased A-cells, reduced B-cells and exocrine fibrosis, quantitative changes in the pancreas in type 2 diabetes. Diabetes Res 1988;9:151–159.
5. Goto Y, Kakizaki M, Masaki N. Spontaneous diabetes produced by selective breeding of normal Wistar rats. Proc Jpn Acad 1975;51:80–85.
6. Goto Y, Suzuki K, Sasaki M, Ono T, Abe S. GK rat as a model of nonobese, non-insulin-dependent diabetes. Selective breeding over 35 generations. In: Shafrir E, Renold AE (eds) Frontiers in Diabetes Research. Lessons from Animal Diabetes II. London: John Libbey, 1988; 301–303.
7. Portha B, Serrades P, Bailbe D, Suzuki K, Goto Y, Giroix M-H. β-cell insensitivity to glucose in the GK rat, a spontaneous nonobese model for type II diabetes. Diabetes 1991;40:486–491.
8. Ostenson C-G, Khan A, Abel-Halim SM, Guenifi A, Suzuki K, Goto Y, Efendic S. Abnormal insulin secretion and glucose metabolism in pancreatic islets from the spontaneously diabetic GK rat. Diabetologia 1993;36:3–8.
9. Giroix MH, Vesco L, Portha B. Functional and metabolic perturbations in isolated pancreatic islets from the GK rat, a genetic model of non-insulin dependent diabetes. Endocrinology 1993;132:815–822.
10. Brownlee M. Glycation products and the pathogenesis of diabetic complications. Diabetes Care 1992;15:1835–1843.
11. Matsuoka T, Kajimoto Y, Watada H, Kaneto H, KishimotoM, Umayama Y, Fujitani Y, Kamada T, Kawamori R, Yamasaki Y. Glycation-dependent, reactive oxygen species-mediated suppression of the insulin gene promoter activity in HIT cells. J Clin Invest 1997;99:144–150.
12. Kaneto H, Fujii J, Myint T, Miyazawa N, Islam KN, Kawasaki Y, Suzuki K, Nakamura M, Tatsumi H, Yamasaki Y, Taniguchi N. Reducing sugars trigger oxidative modification and apoptosis in pancreatic beta-cells by provoking oxidative stress through the glycation reaction. Biochem J 1996;320:855–863.
13. Slack JMW. Developmental biology of the pancreas. Development 1995;121:1569–1580.
14. Teitelman G, Lee JK. Cell lineage analysis of pancreatic islet cell development: glucagon and insulin cell arise from catecholaminergic precursors present in pancreatic duct. Dev Biol 1987; 121:454–466.
15. Teitelman G, Alpert S, Polar JM, Martinez A, Hanahan D. Precursor cells of mouse endocrine pancreas coexpress insulin, glucagon and the neuronal proteins tyrosine hydroxylase and neuropeptide Y, but not pancreatic polypeptide. Development 1993;118:1031–1035.
16. Vinik A, Pittenger G, Rafaeloff R, Rosenberg L, Duguid W. Determinants of pancreatic islet cell mass: a balance between neogenesis and senescence/apoptosis. Diabetes Rev 1996;4: 235–263.
17. Zhu M, Noma Y, Mizuno A, Sano T, Shima K. Poor capacity for proliferation of pancreatic β-cell in Otsuka-Long-Evans-Tokushima fatty rat: a model of spontaneous NIDDM. Diabetes 1996;45:941–946.
18. Reed JC. Bcl-2 and the regulation of programmed cell death. J Cell Biol 1994;124:1–6.

Index of authors

Keyword index